Hidden In Plain Sight 7

Andrew Thomas studied physics in the James Clerk Maxwell Building in Edinburgh University, and received his doctorate from Swansea University in 1992.

His *Hidden In Plain Sight* series of books are science bestsellers.

Also by Andrew Thomas:

Hidden In Plain Sight
The simple link between relativity and quantum mechanics

Hidden In Plain Sight 2
The equation of the universe

Hidden In Plain Sight 3
The secret of time

Hidden In Plain Sight 4
The uncertain universe

Hidden In Plain Sight 5
Atom

Hidden In Plain Sight 6
Why three dimensions?

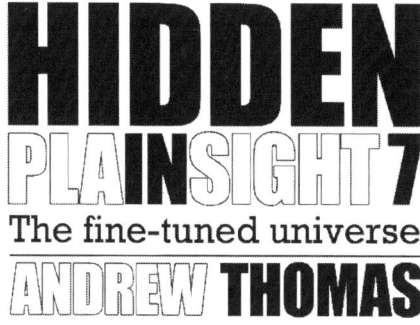

The fine-tuned universe

ANDREW THOMAS

AGGRIEVEDCHIPMUNK.WORDPRESS.COM

Hidden In Plain Sight 7

Copyright © 2017 Andrew D.H. Thomas

All rights reserved.

ISBN-13: 978-1542749671
ISBN-10: 1542749670

CONTENTS

1 The anthropic principle   1
    The six numbers of Martin Rees
    How to predict the winner of every horse race
    The selection effect
    The weak anthropic principle
    The strong anthropic principle

2 Bayes' theorem   25
    Smallpox – or not?
    Putting numbers into the equation
    The usual form of the equation
    How Bayes' theorem won the war
    The controversy over prior probability
    The Bayesian fine-tuning equation
    The probability of life

3 The Fermi paradox   51
    6EQUJ5
    The Fermi paradox
    Exoplanets
    The habitable zone
    Is the Earth special?
    Implications for fine-tuning

4 The cosmological constant 85
   The priest
   The return of the cosmological constant
   Naturalness
   The naturally-flat universe

5 Quantum field theory – part one: 97
Everything is fields
   A brief refresher on quantum mechanics
   The Dirac revolution
   Fock space
   Everything is fields

6 Quantum field theory – part two: 115
The Feynman approach
   The totalitarian principle
   Feynman diagrams
   The fine structure constant
   The Higgs mechanism
   The hierarchy problem

7 The weakness of gravity 143
   The Randall-Sundrum model
   The true strength of gravity
   The laws of physics do not change over time
   The early universe
   Why is gravity so weak?
   The fine-tuned universe

Epilogue 169

PREFACE

I am pleased to say that the subject of this book is the hottest topic in fundamental physics research today. Physics blogs and forums are buzzing with discussions about "fine-tuning" and the related subject of "naturalness".

These questions have taken increased priority after experiments at the Large Hadron Collider did not reveal any of the "Beyond the Standard Model" (BSM) physics which many particle physicists hoped would appear. With little new data to analyse, many particle physicists fear for their jobs. Does this represent the beginning of the end of fundamental physics research?

The LHC discovered the Higgs boson, but the mass of the Higgs boson has emerged as something of a puzzle in its own right. As we shall see later in this book, we have no idea why the Higgs mass is so small. Is the value "natural", or is it "fine-tuned"? Is the universe itself fine-tuned for life?

As you can see, these are seriously big questions. Though I have also included several side-stories to keep you entertained.

Once again, thank you for your support.

Andrew Thomas (hiddeninplainsightbook@gmail.com)
Swansea, UK,
2017

*"As I looked out into the night sky,
at all those infinite stars,
it made me realize how insignificant
they really are."*

- PETER COOK, BRITISH COMEDIAN

1

THE ANTHROPIC PRINCIPLE

In 1973, Brandon Carter, an Australian-born theoretical physicist, had just written a scientific research paper containing an idea which he knew was potentially explosive. Carter, who was studying at Cambridge University at the time, was unsure about the wisdom of publishing such a controversial paper. However, a young Stephen Hawking was intrigued by Carter's work and encouraged him to present his idea to the world.

By one of those strange twists of fate which sometimes occur, 1973 also happened to mark the 500th anniversary of the birth of Nicolaus Copernicus, the great Polish mathematician and astronomer. It was Copernicus who first proposed the theory that the Earth was not the centre of the universe, but, in fact, the Earth orbited the Sun. Previously it had been believed that the Earth held a very special position in the universe. But according to the Copernican principle, the Earth was nothing more than just another mediocre planet, and humanity was nothing more than some unremarkable chemical compound smeared over the face of the planet.

But now, for the first time in 500 years, the science of Brandon Carter was about to challenge Copernicus.

To mark the anniversary of the birth of Copernicus, a conference was being held in the Polish town of Krakow where Copernicus had studied. The timing was perfect for Carter. If Carter wanted to create a big stir in the audience, he certainly managed it. He picked just the right occasion to maximise the impact of his paper.

Carter's important and influential paper is available here:

http://tinyurl.com/carterpaper

Carter agreed with Copernicus that our position in the universe was not special or central in every way. However, Carter realised that this did not mean that our position could not be special in **any** way. Indeed, in some ways it was clear that our position **was** privileged. As an example, the Earth orbits the Sun at a comfortable distance so that its surface is not boiling hot like Venus, or freezing like Neptune. As another example, the strength of the Earth's gravity has the perfect value to retain a thick atmosphere, while not being too strong to crush any complex lifeforms which might emerge. It does seem fairly inarguable that our position – as far as the emergence of life is concerned – **does** appear to be privileged to some extent. Take a trip to Venus if you don't believe me.

But Carter went even further.

Carter had realised that not only was the Earth uncannily suited for life, but also that the laws of physics and the values of the fundamental constants seemed to have forms and values which appeared particularly suited for the emergence of life. If some of the values were only slightly different then it appears that the emergence of life in the universe might well have been impossible. It was almost as if the universe was "fine-tuned" for life.

THE ANTHROPIC PRINCIPLE

As the great physicist John Wheeler said:

> *It is not only that man is adapted to the universe, the universe is adapted to man. Imagine a universe in which one or another of the fundamental dimensionless constants of physics is altered by a few percent one way or another. Man could never come into being in such a universe.*

The six numbers of Martin Rees

It turns out that there is quite an extraordinary number of these "life-friendly" coincidences. Martin Rees has written a book called *Just Six Numbers* in which he considers just six of these remarkable numbers. According to Martin Rees: "The nature of our universe is remarkably sensitive to these numbers. If you can imagine setting up a universe by adjusting six dials, then the tuning must be precise in order to yield a universe that could harbour life."

Here are the six numbers of Martin Rees:

1) The value of N, which is the ratio of the strength of the electromagnetic force to the gravitational force. N has the value approximately equal to 10^{40}, which is 10,000,000,000,000,000,000,000,000,000,000,000,000,000.

   Put simply, this means that the force of gravity is extraordinarily weaker than the electric force. However, unlike the electric force – which can be either attractive or repulsive – gravity is always attractive. So, even though gravity is extremely weak, the attractive force accumulates and becomes dominant for large masses. In fact, gravity becomes so strong that any planet larger than Jupiter would become crushed by gravity – so the Earth is very lucky that gravity is not stronger. Not only would the

Earth be crushed, but all life on Earth would be crushed as well.

Similarly, as gravity becomes stronger for larger masses, this means that stars must necessarily be huge in order to overcome the tremendous outward pressure generated by the fusion reactors in their core. It was the American physicist, Robert Dicke, who first realised that stars are so big because gravity is so weak. If gravity was stronger then stars would be very much smaller, and their lifetimes would then be too short to allow the evolution of complex life.

The strength of gravity will be considered in detail in the final chapter of this book.

2) The value of ε, the Greek letter *epsilon*, which represents the amount of energy released by the nuclear fusion reaction inside stars. The main fusion reaction combines hydrogen atoms to form helium atoms. The nucleus of a hydrogen atom is formed of a single proton, whereas the nucleus of a helium atom is formed out of two protons and two neutrons. But the nucleus of a helium atom weighs only 99.3% as much as the two protons and the two neutrons out of which it is made. The remaining 0.7% is released as energy during the fusion reaction, the energy coming from the loss of that small amount of mass according to $E=mc^2$. Hence, the value of epsilon is 0.7%, or 0.007.

Martin Rees then considers the case if the value of epsilon was 0.006 instead of 0.007. This would be the case if the strong nuclear force holding the protons and neutrons together was slightly weaker. As a result of this weakening of the force, protons could not be bound together during fusion and we would be left with a universe composed entirely of hydrogen. There would be no chemistry in such a universe – and no life.

Rees then considers the alternative situation in which epsilon was 0.008 instead of 0.007. This would represent a strengthening of the strong nuclear force. In that case, all the hydrogen would have disappeared from the universe shortly after the Big Bang, quickly fused into helium. There would then have been no fuel for stars, no water, and no life.

So, according to Martin Rees: "If epsilon were 0.006 or 0.008, we could not exist."

3) The value of Ω, the Greek letter *omega*, which represents the ratio between the total amount of mass and energy in the universe, and the *critical density* of the universe (the formula for the critical density was derived in my second book). The value of omega determines the eventual fate of the universe. If the total mass and energy contained within the universe is too small (less than the critical density), then the force of gravity will be too small to pull the universe back together again, and the universe will expand forever (this situation is called an *open* universe). Conversely, if the total mass and energy contained within the universe is larger than the critical density, the universe will eventually get pulled back together again by gravity (this situation is called a *closed* universe).

Observations of the universe indicate that the value of omega is close to one (indicating a *flat* universe, rather than open or closed). This appears to represent another case of fine-tuning of the value of omega: if the value of omega was much less than one, the universe would have expanded too fast to allow galaxies and stars to condense, and there would be no chance of life. Conversely, if the value of omega was much greater than one, the universe would have collapsed into a "Big Crunch" too quickly for stars to form. According to Martin Rees: "It looks surprising that our universe was initiated with a very finely-tuned impetus, almost exactly enough to balance the decelerating tendency of gravity. It's like sitting at the bottom of a well and throwing a stone up so that it comes to a halt exactly at the top – the required precision is astonishing."

There are, however, a couple of caveats to this apparent fine-tuning coincidence. Firstly, the popular inflation hypothesis predicts a flat universe – without the need for fine-tuning (though it has been suggested that inflation itself requires additional fine-tuning[1]).

Secondly, this entire discussion is based on the behaviour of gravity at the largest of scales: will gravity pull the universe back together, or will gravity be too weak to prevent continual expansion? If our knowledge of gravity at the scale of the entire universe is wrong then it will have to be replaced by a theory of *modified gravity*. In that case, the fine-tuning requirement might be avoided. This possibility will be considered later in this book.

4) The value of Λ, the Greek letter *lambda*, which represents the *cosmological constant*.

We now know that the expansion of the universe is accelerating. This is due to a phenomenon which has been called *dark energy*. Dark energy is often suspected to be the vacuum energy present in empty space. However, our best estimate of the value of the vacuum energy is wildly inaccurate, and it appears that the actual value must be fine-tuned to an incredibly small value.

We will see later how Leonard Susskind has described the apparent fine-tuning of the cosmological constant as a

---

[1] Paul Steinhardt, *The Inflation Debate*,
http://tinyurl.com/PaulSteinhardt

THE ANTHROPIC PRINCIPLE

"cataclysm" for physics. The cosmological constant will be considered in detail in Chapter Four of this book.

5) The value of Q, the size of the initial "ripples" in the structure of the universe which seed the growth of all cosmic structures.

Martin Rees explains the importance of these initial irregularities being sufficiently large: "If the universe had started off completely smooth and uniform, it would have remained so throughout its expansion. It would be cold and dull: no galaxies, therefore no stars, no periodic table, no complexity, certainly no people." However, if the ripples had been too large then massive chunks of matter would have congealed into vast black holes. There seems to be some conflicting opinion over the problems introduced by a large Q, though there seems to be agreement over the necessity that Q should not be too small.

6) The value of D, the number of spatial dimensions of the universe – which is obviously set to three dimensions.

The number of spatial dimensions was the subject of my previous book. As I explained there, life could not exist if D was equal to two or four. However, in my book I came to the conclusion that this number was unlikely to be a fine-tuned coincidence. Any integer value close to one is likely to have a "natural" explanation (we shall be considering *naturalness* again later in this book). There is likely to be a simple mechanism which generates the number of dimensions, such as the simple mechanism which was proposed in my previous book.

With his listing of six arbitrary numbers, Martin Rees presented an analogy of six dials which have to be turned to particular values. As Martin Rees said: "The nature of our universe is remarkably sensitive to these numbers. If you imagine setting up a universe by adjusting six dials, then the tuning must be precise in order to yield a universe that could harbour life."

Here are Martin Rees's six numbers represented as dials:

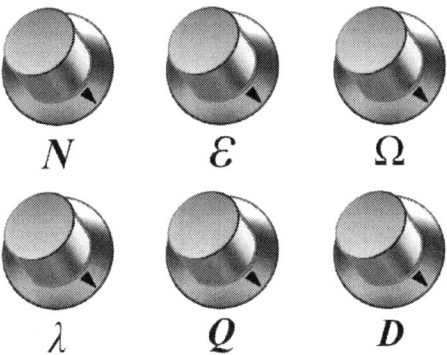

There are more than just six arbitrary numbers, however. The Standard Model of particle physics requires the setting of the value of 25 fundamental constants – which would require a lot of dials![2]

So Brandon Carter's 1973 paper really did cause a stir, the implications of which are still being felt to this day. But what made Carter's paper so controversial was that he also proposed an explanation for the fine-tuning coincidences.

To understand Carter's explanation, we will need to spend a day at the races ...

---

[2] See the Wikipedia page on "Dimensionless Physical Constant": http://tinyurl.com/physicalconstant

THE ANTHROPIC PRINCIPLE

How to predict the winner of every horse race

Derren Brown is one of the most impressive performers currently on British television. His main act is that he is a mentalist, which means he appears to read people's minds. His mind reading act is so impressive that you become convinced he must have special powers. Brown can also get people to perform crazy feats through mind control. For example, he once convinced a group of old age pensioners to commit an art robbery – purely by mind control (the gallery was made aware of the impending theft).

What makes Brown's act so appealing is that he obviously has a good grasp of the science behind his technique, and he mixes scientific and mathematical techniques in with his magic. Sometimes it is difficult to tell where the science ends and the magic begins.

On British television in February 2008, Brown presented what is often regarded as his masterpiece. It was, quite simply, a way of predicting the winner of every horse race.

The show was called *The System*. Here is a description of the system in Brown's own words as he stands in the crowds of Sandown Park Racecourse at the start of the programme:

> *I have developed a guaranteed system for winning at the horses. This system allows me to predict – 24 hours in advance – which horse will win in big, high-profile races. Now, to prove this, six weeks ago I took a woman – a random member of the public – and I told her which horse was going to win in a certain race. It did win. She was intrigued. I then did it again, and again, and again. She started to bet larger and larger amounts of money. Now today, that woman has scraped together every last penny that she can find, and she is risking it all on one final race.*

Ok, to clarify: this was a woman who was genuinely selected as a random member of the public. This was a single mother, a charming lady called Khadisha, who was working two jobs and who Derren Brown thought could most benefit from the winning money. Brown sent her a series of anonymous tips via email for several weeks. Because of Brown's apparent infallibility at picking winners, Khadisha was persuaded to bet large amounts of her own money on high-profile races. On the first race, Khadisha won £28. The second race went to a photo finish and there were a few tense minutes before it was announced that Khadisha's horse had won again. This time Khadisha won £360.

The fifth race proved particularly exciting with the predicted winner lying a long way back in third place going into the final jump fence. However, at the final fence the two leading horses both fell, so Khadisha's horse was then able to stroll through to an easy victory. Khadisha won just under a thousand pounds. Her only regret was that she had not betted more money so that she could have retired on the winnings.

So Derren Brown persuaded Khadisha to raise £4,000 (including £1,000 borrowed from her father) to bet on the sixth and final race. In the final race, the predicted horse won again (after a bit of classic showmanship from Brown) and Khadisha took home £13,000.

The entire video of *The System* is available on YouTube at the following link:

http://tinyurl.com/derrenbrownsystem

If all else fails, the following link to the video should always work:

http://tinyurl.com/derrenbrownbackup

At this point, it might be a good idea to put down this book and go and watch the video, and try to work out how such a remarkable feat could have been achieved. I guess the solution would be called an example of lateral thinking – it is quite ingenious. I can tell you that I managed to work out the secret before Brown explained how he did it at the end of the video. See if you can do the same.

At about the nine-minute mark of the video, Brown announces that he is going to perform another seemingly impossible feat:

> *Part of what makes the System seem so impossible is that it defies our understanding of probability. I'm going to show you something now which is impossible in exactly the same way. I am going to toss a coin, fairly, ten times in a row and have it come up heads every time.*

As Brown said, this was to be filmed under controlled conditions with multiple cameras running continuously which would not cut away, and it was a genuine coin with heads on one side and tails on the other. But then Brown said something more:

> *I want you to watch this (the coin tossing) and try and work out how it can be possible. Because the key to understanding this is the key to understanding the System.*

Indeed, it was after Brown had performed this second seemingly-impossible feat that I worked-out the basic secret of how Brown was going to achieve his feat of predicting the winners of the horse races. Because, while there could be an element of skill in predicting the results of the horse races (by possessing expert knowledge of the horses, the jockeys, and the conditions), there could be no such skill which would help you throw a coin and get ten heads in a row.

There was only one possibility, and that is we were not being told the whole story.

The selection effect

Warning: spoilers ahead.

At about the 29-minute mark of the video, Derren Brown explains how the System works.

He starts by considering a similar example, an example of a person who takes a homeopathic remedy (homeopathic medicine is generally regarded as being unscientific) but finds it to be an effective cure:

> *You might have a viral infection, you might take the homeopathic remedy, you might then feel better and decide, therefore, that it must be an effective cure. The point is, it works for you – what more proof could you need? The trouble is that when these things are tested properly over thousands of people they are shown to have no effect whatsoever.*
>
> *The trap that people fall into is that they do not realise how limiting their own perspective can be. Khadisha believes in the system – she is convinced by it – because she is only looking at it from her own perspective. Now it is time to force a change in perspective, and to look at the bigger picture.*

So why is all this relevant to the theme of this book, trying to decide whether or not the universe is "fine-tuned" for our existence? Well, it is precisely because we can only ever consider the problem from our own perspective – and that can be deceptive. We might interpret our existence as being the result of an impossibly improbable sequence of events, just as it was deceiving Khadisha picking her impossibly improbable sequence of winning horses (Derren Brown calculated that the probability of picking the sequence of winning horses correctly was one in 1.48 billion).

As Derren Brown said: "To work out the System, you need to understand that we can only know what comes from our own limited experience. And our experience can often be very far from the truth."

To explain the secret of the System, Derren Brown then returns to consider his feat of predicting ten correct coin tosses in a row:

> *To predict a run of ten heads in a row, and then make it happen, is hugely unlikely: the chance of it happening is about one in a thousand. However, if you flip a coin thousands of times, and record the results, then somewhere along that line of heads and tails then a line of ten heads is actually very likely to appear.*

So that is how Derren Brown did his coin trick. What we saw – Brown tossing ten heads consecutively – represented only the final minute of what he called "an excruciatingly long day". Brown and his team filmed for over nine hours, with Brown tossing the coin continuously until he threw a consecutive sequence of ten heads. As Brown says: "the impossible became the inevitable".

So only the last minute of Brown's long day was selected for broadcast. But from our very limited viewpoint, it appeared that Brown had simply thrown ten heads in a row. What appeared to be an incredible coincidence was actually an inevitability.

Then Derren Brown explains how the horse racing trick was done. It was based on a similar principle to the coin tossing trick in that we only got to see a very small part of the bigger picture. Initially, 8,000 people were contacted by email – not just Khadisha. Each of those 8,000 people received different tips as to which horse would win the first race. After the race, apologies were sent to all the people whose horse did not win and their bets were refunded. Khadisha happened to be in the group of people who had a winning horse in the first race.

The remaining people were then given different tips on a second race. Again, apologies (and refunds) were sent to the losers, and only the winners (including Khadisha) progressed to the next race.

By the time of the sixth race, there were only six people left, and they were each assigned a different horse in the final race. Of course, one of them **had** to win – the impossible became inevitable. It just happened to be Khadisha who received winning tips for all six races, and it was only Khadisha's story which was broadcast.

So what we have here is a large group of initial data (a large group of people) and only a small sample was taken from that group (a single person). But there was clear bias in the selection process as only the person who won all six races was ever going to be chosen as the representative sample.

Let us now consider another example of selection bias.

Let us imagine we want to determine the percentage of the population who have access to a telephone. The way we might decide to determine this is to conduct a poll: to pick a small sample of the population, see how many of them have a telephone, and then extrapolate that result to determine the percentage of the general population who have telephones. However, the way we decide to select our sample is via a phone poll: we ring up 1,000 people at random. I am sure you can anticipate the problem here. Out of that 1,000 people, we find that 100% of them have telephones. We then extrapolate that result to decide that everyone in the general population has a telephone. But, of course, there was clear bias in the way we selected our sample: only people who had a telephone in the first place could have been contacted via our phone poll. So of course it appeared that everyone had a phone.

When there is bias in the selection of a sample from a larger population, the misleading result is called a *selection effect*. It has been suggested that the apparent life-friendly coincidences in our universes could be a result of selection effects – in which case the apparent fine-tuning of the universe would be nothing more than an illusion. Let us now consider this suggestion.

The weak anthropic principle

In our previous discussion of selection effects, we have seen that any form of selection bias in a sample of data can lead to misleading conclusions being made. In particular, it can appear that uncanny coincidences have occurred, when, in fact, no such coincidences exist. With this in mind, when we consider the nature of the universe – to discuss whether or not it is fine-tuned – we can see that our existence as a form of intelligent life acts as a form of selection bias.

This is fairly obvious, after all, if conditions in our particular part of the universe were not amenable to the emergence of life then we would not be here to observe the life-friendliness of the universe – we simply would not exist. So, just like the previous example of the phone poll, if we ask the opinions of a privileged few people (people with telephones, or intelligent life living in comfortable regions of the universe) then you are going to get a false impression. On the basis of the responses, the universe is going to seem like a privileged environment, uncannily friendly and comfortable for life (with plenty of telephones as well).

So life is only ever going to be found in regions of the universe which are amenable to life. And any intelligent lifeform is then going to interpret the universe as being life-friendly. As Derren Brown pointed out, we can only ever consider the problem from our own perspective – and that can be deceptive. This tendency of intelligent life to interpret the universe as life-friendly is a selection effect and is fairly obvious. It should therefore be regarded as a tautology or truism.

In his book about the anthropic principle called *The Goldilocks Enigma* (also known as *Cosmic Jackpot*), Paul Davies makes this point: "Nobody can be at all surprised by this simple tautology. It merely says that observers will find themselves located only where life can exist. It could hardly be otherwise."

This fairly obvious principle is called the *anthropic principle*. The anthropic principle was presented for the first time (and given its name) in Brandon Carter's seminal 1973 paper, which was linked earlier but I will repeat here:

http://tinyurl.com/carterpaper

Specifically, this simple version of the anthropic principle which explains life-friendly coincidences as selection effects is called the *weak anthropic principle*.

As an example of the weak anthropic principle, we might look at the surface of Venus through a telescope and see frightening conditions: towering volcanoes over a lava surface, with surface temperatures over 700 degrees Celsius as heat is trapped by its 95% carbon dioxide atmosphere (it is actually the worst-case scenario for the greenhouse effect). On the plus side, it does rain occasionally – but the rain is composed of sulphuric acid.

Now, when we observe the surface of Venus, our naive reaction might be to think: "What an awful place! Isn't it lucky we don't live there? It is surely an uncannily fortunate coincidence that we live on comfortable Earth and not horrific Venus."

But, of course, the weak anthropic principle explains this apparent lucky coincidence as a mere selection effect: intelligent life is only ever going to be found in life-friendly environments. It is no coincidence we are based on Earth and not Venus: human beings could never have evolved on the surface of Venus.

The weak anthropic principle states that if conditions vary with location throughout the universe, then life will only emerge in locations in which those conditions are amenable for life. That selection effect might then result in intelligent life naively interpreting the universe as being uncannily life-friendly, i.e., "fine-tuned".

So the weak anthropic principle is really fairly obvious, and could be regarded as common sense. Unfortunately, like any fairly-obvious common sense idea, the weak anthropic principle does not provide us with any particularly powerful insights about life or the universe. The weak anthropic principle should be regarded as a truism, and the Google definition of "truism" is: "a statement that is obviously true and says nothing new or interesting" – which is a perfect description of the weak anthropic principle. It is called "weak" for a very good reason. All the weak anthropic principle really does is act as a warning that we should be careful not to be fooled by some fairly obvious selection effects.

In particular, while the weak anthropic can explain some very simple naive fine-tuning problems (e.g., "Why is the surface of the Earth just the right temperature for life?") it cannot solve the fine-tuning problems we are really interested in. To be precise, the weak anthropic principle cannot be used to explain why the laws of physics and the values of the fundamental physical constants are apparently life-friendly. If we want to use the anthropic principle to explain these deeper coincidences, then we are going to have to take a deep breath and consider a much more controversial and contentious principle.

We are going to have to consider the strong anthropic principle.

The strong anthropic principle

In the previous discussion of the weak anthropic principle we have considered properties which are important to life (such as temperature, abundance of chemical elements) which vary with location throughout the universe. In that case, our interpretation of conditions as being life-friendly can be explained as a simple selection effect. So, if conditions vary with time or space, the weak anthropic principle can explain the apparent life-friendliness of our environment.

However, in this book we are interested in whether or not the laws of physics and the values of the physical constants are fine-tuned for life, and it appears that the laws of physics and the values of the physical constants have not changed with time and do not change with location. If we observe deepest space with our most powerful telescopes, the same laws of physics seem to apply in deepest space as in our local laboratory. Also, the value of the fundamental constants are not believed to have altered over time.

So the weak anthropic principle cannot explain the apparent fine-tuning of the universe. But there is a more powerful version of the anthropic principle which attempts to do just that.

The *strong anthropic principle* (again, described for the first time in Brandon Carter's 1973 paper) considers the situation in which the laws of physics and the value of the constants **do** vary with location. In that case, once again, the life-friendliness of the universe can be explained merely as a series of selection effects. But how can the laws of physics possibly change with location? This is only possible if we take a highly-speculative leap into the unknown.

It is only possible for the laws of physics to change with location if we imagine there to be other "bubble universes" located outside our own observable universe. The laws of physics, and the value of the fundamental constants, might then be set to different values in each universe.

The following diagram shows our universe surrounded by various other bubble universes. The total collection of all these bubble universes is called the *multiverse*. So, as we vary our location throughout this multiverse, we find the laws of physics varying:

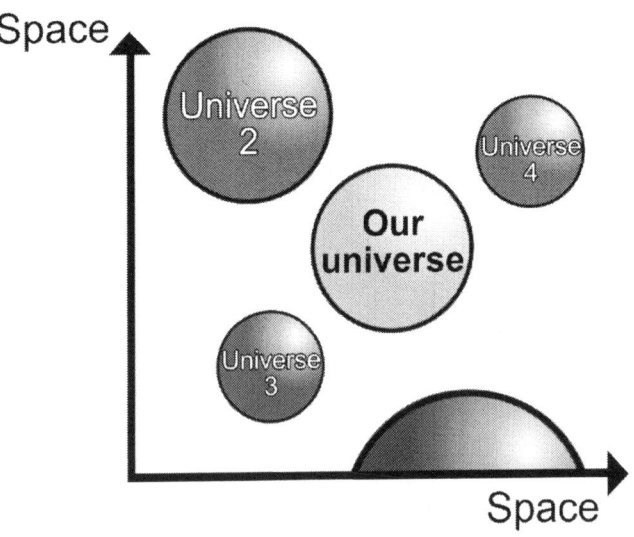

This is called the *multiverse hypothesis*.

If the multiverse really exists then the strong anthropic principle can be used to explain anthropic coincidences as mere selection effects.

What is more, if there really is a multiverse then, as Brandon Carter suggested for the first time in his 1973 paper, we can use it to provide an explanation as to why the laws of physics take the form they do. The laws of physics would take the form we observe simply because we happen to inhabit the particular universe in which the laws of physics and the fundamental constants are set to those particular values. If we happened to inhabit a different universe, we would observe the laws of physics and the values of the constants set to different random values.

The strong anthropic principle is finding a new-found popularity at the moment due to its apparently unlimited explanatory potential: it can explain anything – the form of the laws of physics can be explained as random selection effects. This idea was first proposed in Section Five of Brandon Carter's paper which I linked earlier (he refers to the multiverse as a "world-ensemble"):

> *It is of course philosophically possible – as a last resort, when no stronger physical argument is available – to promote a **prediction** based on the strong anthropic principle to the status of an **explanation** by thinking in terms of a "world-ensemble".*

This idea first presented by Brandon Carter in 1973 has become highly influential over the last ten years. As progress in fundamental physics has stalled, the strong anthropic principle appears to present an easy and attractive means of "explaining everything". I considered anthropic reasoning in detail in my fourth book when I repeated the common criticism that a theory which predicts everything predicts nothing. I returned to this theme in my previous book when I suggested that we should consider anthropic theories to be "theories of last resort" – just as Brandon Carter suggested in his previous quote.

What is more, the whole multiverse hypothesis is frequently criticised as being unscientific as it cannot be tested or disproved.

At the end of his paper, Brandon Carter presented his own concerns about the use of the strong anthropic principle to determine the values of the physical constants:

> *The acceptability of predictions of this kind as explanations depends on one's attitude to the world-ensemble concept. The idea that there may exist many universes, of which only one can be known to us, may at first sight seem philosophically undesirable.*
>
> *I would personally be happier with explanations of the values of the fundamental coupling constants etc. based on a deeper mathematical structure.*

Paul Davies agrees with this viewpoint in his book *The Goldilocks Enigma*, stressing that the promotion of the strong anthropic principle undermines the efforts of physicists who are attempting to develop conventional analytical methods for understanding the universe:

> *For those theoretical physicists hard at work trying to formulate a unique final theory, the multiverse comes across as a cheap way out. Transforming cosmology into a messy environmental science looks a shabby let-down when set alongside the inspiring magnificence of a unique final theory that would explain everything. Randomness plus observer selection strikes many physicists as an ugly and impoverished explanation compared with an overarching mathematical theory.*

I hope in this discussion I have managed to convey the impression that the anthropic principle is really not a very useful tool for explaining anything: the weak anthropic principle is too weak, and the strong anthropic principle is too speculative and too unscientific. What is more, as we have discussed, anthropic reasoning based on the strong anthropic principle threatens to undermine the whole process of doing good physics.

In short, the anthropic principle is bad news all round.

But the anthropic principle has certainly had a wide influence. In my experience, there is a tendency in the educated wider public (and many physicists) to dismiss **all** life-friendly coincidences as merely selection effects, influenced by the weak or strong anthropic principle. Certainly my friends I have talked to about this subject seemed to think like that – they seemed to be under the impression that the fine-tuning problem had already been explained as a simple selection effect, perhaps unknowingly invoking the anthropic principle. However, as this discussion has shown, it is unjustified to dismiss these coincidences as selection effects. If we try to explain the apparent fine-tuning of the laws of physics and the fundamental constants as mere selection effects then we have to introduce highly-speculative and controversial ideas, such as the multiverse hypothesis.

Instead, the message which I will be presenting in this book is that we need to follow a different, more scientific approach to explaining the life-friendly coincidences in the universe. Basically, instead of jumping to embrace simple solutions, we need to acknowledge that these problems are difficult – and rise to the challenge.

2

BAYES' THEOREM

Essentially, the question of whether or not the universe is fine-tuned comes down to a question of probability. If it is the case that the universe is fine-tuned, then that is equivalent to saying that the fundamental constants have surprising, improbable values. In that case, if the values are improbable, then we would not normally expect to encounter them by pure chance – something strange is happening. Conversely, if the universe is **not** fine-tuned, then that is equivalent to discovering that the fundamental constants having highly-probable, highly-likely values.

So, the fine-tuning question is really a question about probability, and in this chapter we will be considering the science of mathematical probability. We will be seeing if the universe really is – as has been suggested – "Unlikely. Very unlikely. Deeply, shockingly unlikely."[3]

---

[3] Why is There Life? *Discover Magazine*, http://tinyurl.com/unlikelyuniverse

As human knowledge advances relentlessly, it is very comforting to imagine that we will eventually know everything – though that might prove to be an impossible goal. There might always be some pockets of ignorance. The casinos of Las Vegas rely on that ignorance for their income. For example, we are completely ignorant of the outcome when we roll a pair of dice. Will we roll a seven, or ten, or five? We are simply ignorant. However, if we roll the dice multiple times then we can detect a pattern in the results, and though we still cannot predict the result of an individual throw, we can predict the probability of how many times a certain result will appear over time. For example, if we throw the dice one hundred times, we can calculate how many times we would expect to roll a seven.

Probability is expressed as a number between zero and one, with a probability of zero indicating a result is impossible, and a probability of one indicating a result is certain.

So mathematical probability gives us a way of moving from an effect (the throwing of the dice, for example) to predicting a likely cause (the result of the throw). Hence, probability moves from cause to effect. Probability allows us to deal with our ignorance when we have knowledge about the cause, but we are ignorant about the result.

But what about the reverse situation? What happens if we have ignorance about the cause, but we have knowledge about the result? For example, if we are presented with a series of numbers (one, thirteen, seven, etc.) resulting from a series of dice throws, can we tell **purely from the results** whether or not the dice were fair, or if they were loaded towards a particular result? In other words, instead of moving from cause to effect (as in probability) can we move in a backward direction from the effects backwards to consider the cause? This is called the *inverse probability problem*.

The solution to the inverse probability problem is far from intuitively obvious. Consider an example relating gun homicide to gun ownership. If we consider police records of every murderer who committed their crime using a gun, we might find that in 95% of those cases a gun was later found in the house of the murderer. So there is a probability of 0.95 that a murderer will have a gun in their house.

Now let us consider the inverse probability: what is the probability that if we find a gun in a house, then the owner will be a murderer? Our first thought might be that the probability in the reverse direction would be the same as in the forward direction, and therefore suggest that there is a 95% chance that the owner will be a murderer. But clearly, this is nonsense: gun ownership is widespread in America, and only a tiny fraction of those owners will ever turn out to be murderers.

So how do we calculate the inverse probability correctly? The solution to the inverse probability problem is provided by *Bayes' theorem*, named after the 18th century English statistician Thomas Bayes.

Bayes' theorem explains how we can calculate probability in the backward direction, from effect to cause. We can use Bayes' theorem to calculate a probability that the cause has a certain form: for example, are the dice loaded or not? Is the gun owner likely to be a murderer? By considering the result which we can see – the *evidence* – we can infer the nature of the cause. In our previous examples, we would consider the evidence to be the results of the dice throws, or whether or not a gun has been found in a house.

Bayes' theorem is applicable to many different areas of life – not just science. It is this widespread use of Bayes' theorem which has made it something of a modern sensation, as we shall see later in this chapter.

Smallpox – or not?

The best way to understand Bayes' theorem is to consider an example. In his tutorial book of Bayes' theorem, James Stone presents an example involving a nasty case of smallpox. Let us consider the example.

One morning, you wake up and discover your face is covered in spots. You go to the doctor, and the doctor tells you that 90% of people with smallpox are found to have spots on their face. This news terrifies you, because smallpox is a potentially lethal disease. However, you swiftly realise that this is a strange thing for a doctor to say: the doctor is giving you a probability in the forward direction – from cause (smallpox) to effect (spots). This is no use to you because you want the doctor to give you a probability in the inverse direction. You want the doctor to examine the evidence (the spots) and tell you the probability that you have smallpox.

However, as we saw in the earlier example about gun ownership, it is not obvious how to calculate probability in the reverse direction: just because 90% of people with smallpox have spots, it most certainly does not mean that 90% of people with spots have smallpox.

So how can we use Bayes' theorem to calculate the inverse probability, thus determining the probability that you have smallpox?

BAYES' THEOREM

Well, in order to use Bayes' theorem, the first thing we need to have is some sort of initial belief about the "way things are" – a belief about the state of the world. We call this our *hypothesis*. In the smallpox example, our hypothesis might be "The patient has smallpox". The probability that our hypothesis is true is initially generated before we consider any evidence (before we examine the patient and see the spots). However, we already know that smallpox is extremely rare in the general population, so we would expect the initial probability that our hypothesis is true – in the absence of any supporting evidence – to be extremely small.

Bayes' theorem then tells us how we should update our belief in our hypothesis on the basis of new evidence which becomes available.

So how can we use Bayes' theorem to update our confidence in our hypothesis? Well, we have to examine the evidence: we see the spots on the patient. It would then seem to make sense that we should increase our confidence in our hypothesis (the probability that our hypothesis is correct) if it is highly likely that a patient with smallpox will have spots. In the following equation this value is called the "Probability of the evidence appearing if the hypothesis is true". In other words, it is the probability of spots appearing if the patient has smallpox. In the equation, we multiply our initial probability by this term.

We are almost there, but not quite. We have to take another factor into account. We have to account for the general tendency for spots to appear in the population. If spots are common generally (for less serious diseases such as chickenpox and acne), then it would clearly weaken our hypothesis that the patient has smallpox. So we have to divide by this term: the "Probability of the evidence appearing in general". So we end up with the following equation:

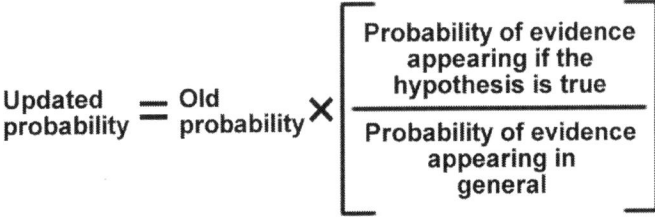

And that is the famous Bayes' theorem!

It really is surprisingly simple, and we have generated the theorem purely by an intuitive approach. If you consider the term in the square brackets in the equation, you will see it will be greater than one if the top half of the fraction has a greater value than the bottom half of the fraction. You will see that this will be the case if the evidence is more likely to appear if the hypothesis is true than in the general case. Hence, in that situation, your updated probability (your new belief in your hypothesis) will be greater than your old belief in your hypothesis: finding the evidence has increased your belief in your hypothesis.

But even though it is extremely simple, the implications of Bayes' theorem are extraordinary. In fact, it has been said that "It does not seem an exaggeration to consider Bayes' theorem as significant in our times as the Darwinian theory of natural selection."[4]

In this chapter we will be discovering how this incredibly simple equation is not only responsible for filtering junk email into our spam folders – but it might also have won the Second World War.

---

[4] Peter Pesic, *Times Literary Supplement*

Putting numbers into the equation

In order to see how valuable Bayes' theorem can be in updating our beliefs, let us put some numbers into the equation. I will be using the same values which James Stone used in his smallpox example.

Firstly, as described earlier, our initial belief that the patient has smallpox is generated before we examine the evidence – before we see that the patient has spots. So all we can do at this early stage is to treat the patient like any other person and set this initial probability that the patient has smallpox equal to the probability that anyone in the general population has smallpox. Therefore, we should consider the public health statistics to determine the probability that anyone in the general population has smallpox (this initial knowledge – before we even start to use Bayes' theorem – is called the *prior probability*). Because smallpox is rare, this is clearly going to be a small probability. According to James Stone: "Public health statistics may inform us that the prevalence of smallpox in the general population is 0.001."

As described earlier, the next term we have to consider is the "Probability of the evidence appearing if the hypothesis is true". In other words, it is the probability of spots appearing if the patient has smallpox. The doctor has already told us that 90% of people with smallpox have spots, so we set this value equal to 0.9.

Finally, as discussed earlier, the last term we have to consider is the "Probability of the evidence appearing in general", in other words, the general tendency for spots to appear in the population (including less serious diseases). This term acts to undermine faith in our hypothesis, so we have to divide by this term. James Stone uses a value of 0.081 for this term.

Let us put these numbers into the previous formula. We find that the final probability that the patient has smallpox (the "updated probability" generated by Bayes' theorem) is equal to:

$$0.001 \times \left[\frac{0.9}{0.081}\right]$$

which, if you calculate the result, is equal to 0.011. This is clearly very much smaller than the value of 90% which the doctor first suggested, so you are greatly relieved.

Bayes' theorem might not quite have saved your life, but you will sleep happier tonight.

The usual form of the equation

Before we move on to consider the interesting uses of Bayes' theorem, let us briefly consider the form in which the theorem is usually presented. Bayes' theorem is usually presented as an equation in a short-hand form which uses very few characters:

$$P(H \mid E) = \frac{P(H) \times P(E \mid H)}{P(E)}$$

I would say this form is less descriptive and more difficult to understand intuitively than the version I presented earlier. However, just for the record, this is what the various terms of this equation mean:

- P(H) is the "old probability" that the hypothesis, H, is true. This is the prior probability based on existing information.

- P(E|H) is called a *conditional probability*. It is what I have called the "Probability of the evidence appearing if the hypothesis is true".

- P(E) is what I have called the "Probability of the evidence appearing in general".

- P(H|E) is called the *posterior probability*, and is the updated probability that the hypothesis is true.

This is a form of the equation which I find very hard to remember – there are simply too many "E"s and "H"s – whereas I can remember the longer, more descriptive, version of the equation I presented earlier.

How Bayes' theorem won the war

Throughout the Second World War, Germany relied on an encryption method to encode orders for its army, air force, and navy. The German *blitzkrieg* high-speed attack, for example, was heavily-reliant on coded radio messages to coordinate artillery, airplanes, and tanks. Most crucially, German U-boats relied on encrypted messages in order to learn the positions of Allied merchant shipping.

The encryption method was highly-sophisticated and extremely difficult to break. Messages were encoded by a device called an *Enigma machine*, which resembled a typewriter encased in a wooden box. Above the typewriter keyboard was a lampboard with a light for each letter of the alphabet.

The following photograph shows an Enigma machine in use in 1943:

There were three wheels inside an Enigma machine. Each time the operator typed a key, one of the three wheels advanced a notch. The typed letter was enciphered, and the resultant encoded letter was illuminated on the lampboard. The complete message was then transmitted in Morse code.

The following diagram shows the three Enigma wheels. At the start of each day, the wheels were initialised by setting the top letter of each wheel according to a codebook. For example, if the codebook stated that the day's keyword was "EJN" then the top letters on each wheel would be set according to the following photograph:

The task for the codebreakers was therefore to determine the Enigma wheel settings for each day. They only had 24 hours to do this as the settings changed at midnight to the next day's settings.

The Enigma code was a *substitution cipher*, in which a single letter in a message (for example, the letter "A") was replaced by another letter (for example, the letter "G"). The following diagram explains how the Enigma wheels operated to perform this encoding:

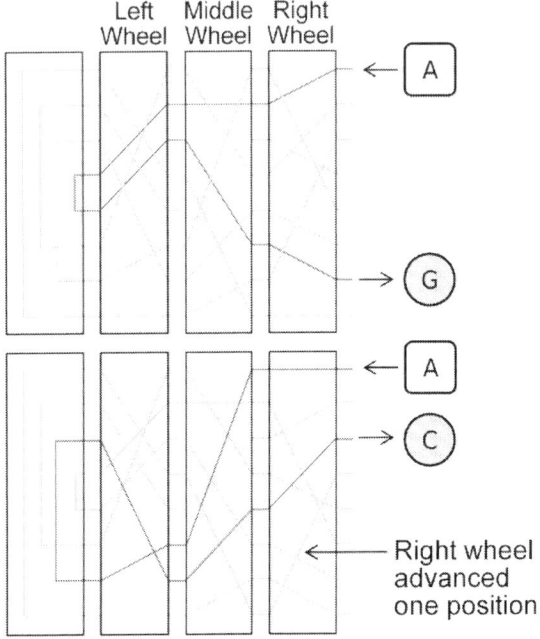

In the diagram, you will see the three Enigma wheels: left, middle, and right. In the top half of the diagram, you will see the letter "A" entering the wheels (on the right), tracing a path through the three wheels (the solid black line), and emerging as the encoded letter "G".

However, the next time the letter "A" was encoded in a message, the right wheel would have advanced one position. So you can see that the letter "A" would then be replaced by a different letter: in this case, the letter "C". Because this made it hard to recognise common sequences of letters, this made the code much more difficult to break.

Another advantage of this structure of the Enigma machine is that the same settings allowed a single machine to act as both an encoder and a decoder. For decoding, the path is just retraced through the wheels in the opposite direction. If you consider the top half of the previous diagram, you will see that entering the letter "G" again will be traced back to the original letter "A". Though the Germans were very attracted to this feature of the machine, it actually introduced a weakness into the coding technique. This is because you will see that this mechanism never lets the same letter be encoded into itself: the letter "A" can never be encoded into the letter "A", for example. This provided a vital clue to the codebreakers.

The British codebreakers were based in Bletchley Park, a Victorian mansion located 50 miles northwest of London. I recently took a trip to Bletchley Park. The complex is no longer a working military facility, but instead the whole area has been turned into a fascinating museum. But this is no theme park – this is real history.

Here I am standing in front of the mansion:

The team were led by the brilliant computing pioneer Alan Turing. Turing realised that the Enigma code could be broken by employing Bayesian techniques. The process of coding a message is from cause-to-effect (the Enigma machine produces a coded message), whereas the process of decoding the message is in the reverse direction from effect-to-cause (analysing the "evidence": the many thousands of encoded messages). Hence, a message could be decoded by using Bayesian techniques and solving the inverse probability problem.

Sharon McGrayne has written a book about the history of Bayes' theorem called *The Theory That Would Not Die*. The book contains a detailed account of the effort to use Bayes' theorem to decrypt the Enigma code. According to McGrayne: "Finding the Enigma settings that had encoded a particular message was a classic problem in the inverse probability of causes."

Once Turing had decided to use Bayes' theorem, it was Bayes' theorem which determined the radically new decoding method which had to be used. Remember, Bayes' theorem operates by considering the evidence in order to decide if a particular hypothesis is correct. In the case of the Enigma decoding, the sheer weight of evidence (all major German communications) had to be analysed in order to find the likeliest hypothesis (one of 159 million million million possible settings of an Enigma machine). This meant that manual decoding was infeasible. Instead, a new decoding technique would be required, capable of handling such a huge amount of information in a short time. Alan Turing realised that an automated system would be required. As David Leavitt says in his biography of Turing, what was required was "a machine built for the specific purpose of defeating another machine."

To achieve this goal, Alan Turing and the team at Bletchley Park – including Tommy Flowers and Max Newman – used electronic equipment for the first time to break codes. In doing so, they created the first programmable electronic digital computer, which was called Colossus.

Instead of modern transistors, Colossus used 1,500 glass valves (vacuum tubes) to switch current on and off (to signify binary zeroes and ones). The computer was so fast that sometimes Bletchley Park were reading the deciphered messages even before their German counterparts. The power of Colossus was so great that it could break the stronger and more sophisticated Lorenz code which was used by Adolf Hitler himself.

Here is a photograph of me inspecting the Colossus rebuild at Bletchley Park. It was the world's first electronic digital computer. Just wow:

Looking at that photograph, it is almost beyond comprehension that it was built in 1944. It looks like a modern server farm! But it's all valves!

You can see more of my super Bletchley Park photographs in my Bletchley Park album:

http://tinyurl.com/andrewbletchley

Later in 1944, Colossus decoded an encrypted message from Hitler to Field Marshal Erwin Rommel, who was his commanding officer in Northern France. Hitler ordered

Rommel not to move his troops for five days if there was an invasion in Normandy. This was because Hitler believed a Normandy invasion would be intended as just a diversion from the true invasion force which would attack the ports along the English Channel. When General Eisenhower – the Supreme Commander of the Allied Forces in Europe – received the decoded message from Colossus, he announced "We go tomorrow". That was to be the morning of June 6th 1944, which was to become better known as D-Day.

Eisenhower later estimated that the Bletchley Park codebreakers shortened the war by at least two years.

The story of Alan Turing's attempt to break the Enigma code was dramatised in the movie *The Imitation Game*. The movie featured a splendid Oscar-nominated performance by Benedict Cumberbatch as Alan Turing. Although the film received some criticism over a few minor inaccuracies, the plot is generally accurate and presents a gripping account of the work at Bletchley Park. The film was a major critical and commercial success, making over $200 million at the box-office.

If you ever find yourself in London, I can recommend Bletchley Park as a great day out. Bletchley Station is just a forty-minute train journey from Euston Station. Walking out of Bletchley Station, turn right, and Bletchley Park is just 50 yards up the road.

The controversy over prior probability

Bayes' theorem now plays a huge role in all our lives – even though most people are completely unaware of it. Remember, Bayes' theorem works by updating the probability that a hypothesis is correct on the basis of received evidence. Hence, Bayes' theorem can act as a form of classification, assigning objects to groups on the basis of that evidence. Here are just a few examples of Bayes' theorem working in this way:

- It is Bayes' theorem which filters our junk email into spam folders.

- It is Bayes' theorem which controls the (sometimes very annoying) autocorrect feature when we type on a mobile phone, selecting the best fit from a range of possible words.

- It is Bayes' theorem which selects adverts which appear on web pages, selected to be of interest to us (the "evidence" controlling the selection process controversially being obtained from our browsing history).

- There is software based on Bayes' theorem which can automatically process your CV and decide whether or not to call you for a job interview.

- It was Bayes' theorem which first connected lung cancer with smoking.

- Richard Feynman used Bayesian techniques to determine that the O-rings were the most likely cause of the *Challenger* space shuttle disaster.

- During the Cold War, the U.S. military think-tank RAND used Bayesian techniques to calculate the probability of a thermonuclear device exploding accidentally (answer: it's a worryingly large probability).

- The designers of Google's self-driving car are using Bayes' theorem to recognise objects and make decisions.

- Perhaps most impressive of all, you can even see Bayes' theorem on Sheldon Cooper's whiteboard in an episode of *The Big Bang Theory*.[5]

In fact, in almost every place where a decision must be made on the basis of statistical data, you will now find Bayes' theorem being used.

Bayes' theorem has proved itself to be so useful and permeates so many aspects of our lives that it may be surprising to hear that until quite recently – the second half of the twentieth century – Bayes' theorem was treated with mistrust and even revulsion by most statisticians. In order to understand the problem, let us remind ourselves how Bayes' theorem is defined:

---

[5] You can see Bayes' theorem on Sheldon's whiteboard at the start of the episode called *The Cruciferous Vegetable Amplification*.

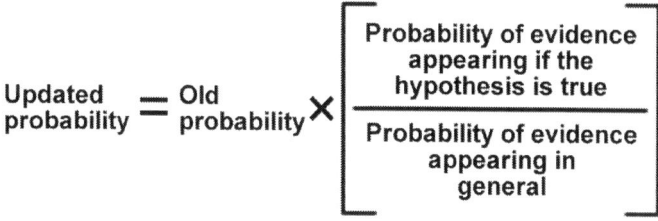

The "Old probability" in this equation is a crucial – and controversial – term. This is the probability which is set before we examine any evidence and before we perform the Bayes' theorem calculation. This probability is the prior probability of the hypothesis being true (often called simply the *prior*). This has proven to be a very controversial term since Bayes' theorem was first proposed, and is the reason why the theorem has been met with so much resistance from statisticians.

If we have access to a large volume of existing data then we can obtain an accurate value of the prior probability with confidence. For example, in the smallpox example considered earlier, it was explained how the prior probability could be calculated from the public health statistics, informing us that the prevalence of smallpox in the general population is 0.001.

However, if we only have very limited access to existing data – or no relevant data at all – there are still other ways to obtain a value for prior probability. Specifically, human experience and judgement can be used to generate a value for prior probability. Though this might appear to be very subjective, it is still valuable information which we should not ignore in our decision process.

So sometimes the value which is used for the prior probability might appear to be no more than an educated guess. Indeed, Thomas Bayes originally referred to this term as a "guess" when he proposed his theorem. This is often called the initial *belief* that the hypothesis is true.

And now perhaps you can see why statisticians have been so unimpressed with Bayes' theorem: what place is there for "belief" in mathematics? As Sharon McGrayne says in her book: "The heart of the Bayesian controversy is the fact that, when only a few data are involved, the outcome of a Bayesian computation depends on prior opinion. In such a case, Bayes' theorem can lead to a subjective, rather than an objective, assessment of a situation."

However, a wonderful feature of Bayes' theorem is that it really doesn't matter which method we employ to generate our prior – as long as we have access to plenty of evidence. As each additional piece of evidence appears and is processed by Bayes' theorem, then an updated – and more accurate – probability of the hypothesis being true is generated. As long as there is plenty of evidence then the final probability is guaranteed to converge on an accurate value. So as each piece of evidence is added, the importance of the value of the initial prior becomes progressively reduced. As Sharon McGrayne says in her book: "with large samples the prior did not matter".

In her book, McGrayne further explores how this tendency of Bayes' theorem to close in on the truth explains how scientists with different initial opinions manage to eventually come to agreement over which hypothesis is correct. McGrayne presents a quote from the statistician Jimmie Savage:

> *If prior opinions can differ from one researcher to the next, what happens to scientific objectivity in data analysis? Well, as the amount of data increases, subjectivists move into agreement, the way scientists come to a consensus as evidence accumulates about, say, the greenhouse effect or about cigarettes being the leading cause of lung cancer. When they have little*

data, scientists disagree and are subjectivists; when they have piles of data, they agree and become objectivists. That's the way science is done.

So we can see how Bayes' theorem closely matches the scientific method, and even matches our individual thought processes when it comes to examining evidence and making difficult decisions. Because such a simple rule has such wide application, Bayes' theorem is almost treated less as a simple tool and more as a guiding philosophy for life by its passionate advocates.

The Bayesian fine-tuning equation

So now let us return to the main subject of this book, and ask: "How can we use Bayes' theorem to determine if the universe is fine-tuned or not?" To determine this, we have to evaluate the probability that the hypothesis – that "the universe is fine-tuned" – is true. Bayes' theorem requires us to obtain evidence to support our hypothesis, and, in this case, the evidence which we observe is the existence of life in our universe.

So let us update our definition of Bayes' theorem (which was presented earlier) with these new values for the hypothesis and the evidence:

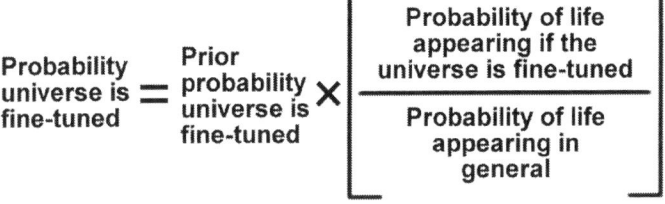

I am going to call this equation the "Bayesian fine-tuning equation".

On the right-hand side of this new equation, the term "Prior probability universe is fine-tuned" represents the prior probability we have just discussed in the previous section. It is this value which is going to be updated by Bayes' theorem to generate an updated probability that the universe is fine-tuned (on the left-hand side of the equation). The available methods for arriving at a prior probability were discussed in the previous section. In the absence of pre-existing data, the value of this term is going to be heavily-reliant on the experience, knowledge, and intuition of physicists.

The question to be asked to generate the prior probability would be: "On the basis of our existing knowledge as physicists, do we currently think the universe is fine-tuned or not?" I would suspect the majority of physicists would currently respond in the negative, though there would also be quite a few who would take the opposing view. Fortunately, as discussed in the previous section, Bayes' theorem is perfectly capable of handling incomplete and contradictory subjective opinions. As the amount of received data increases, the Bayesian fine-tuning equation would update its probability that the universe is fine-tuned, moving toward the correct hypothesis (either "Yes" or "No"). The tendency would then be for scientists to move into agreement about the correct hypothesis – and that is how the Bayesian scientific method works.

So it would appear that the value of the selected prior probability is not crucial. What **is** crucial is the value of the fraction contained in the square brackets, so let us consider it next.

The probability of life

Let us now consider the two terms inside the square brackets.

On the top of the fraction in the square brackets is the term "Probability of life appearing if the universe is fine-tuned". I think we would have to consider this as being a fairly large value: if the universe was fine-tuned for life, creating an environment highly-amenable to the emergence of life, we would surely expect life to spontaneously emerge somewhere in that vast universe – even if it takes billions of years to emerge. So it appears reasonable to set that term to a large value.

But the bottom term of the fraction is not so easy to calculate. And this is a term which is absolutely crucial to our analysis.

You will see that the bottom of the fraction in the square brackets in the Bayesian fine-tuning equation is the term: "Probability of life appearing in general". This represents the probability of life emerging in all cases – including situations where the hypothesis is not true. So this term represents the probability of life emerging in all possible universes – even in universes which are not fine-tuned.

In order to calculate this probability of life, it appears we would have to consider whether or not we would expect life to emerge in any arbitrary universe, even if the physical constants were set to random values in that universe. Maybe the resultant form of life would be quite unrecognisable to us, not resembling human life at all. Basically, the question being asked here is "Does life emerge easily?" or, equivalently "Would we expect to find life in all possible universes?"

If life is common, and emerges easily in all universes, the value inside the square brackets will be small, and so the corresponding probability that the universe is fine-tuned will also be small. This makes sense: if life is robust and emerges freely, then it has no need of additional helpful fine-tuning. Conversely, if life does not emerge easily, the value inside the square brackets will be large, and the probability that the universe is fine-tuned will be high. Again, this makes sense: if life is delicate and rare then it needs all the help it can get from fine-tuning in order to emerge.

So it appears that all we have to do is determine the probability of life emerging in general, and then we will be able to instantly determine whether or not the universe is fine-tuned.

Unfortunately, it is not at all easy to obtain an accurate value for the probability of life emerging. Ideally, we would need to examine a set of different, independent universes, with the fundamental constants set to different random values in each universe, and then see how often life emerges in those varied universes. Well, of course, this is something we can't do: we don't have access to a convenient set of universes on which we can experiment. It appears our cause is hopeless.

However, perhaps there is a second-best option. While we don't have a set of universes with different conditions, there are regions of our own virtually-infinite universe with wildly-different conditions. How likely is life to emerge in these extraterrestrial breeding grounds?

So, let us now consider what we know about the possibility of extraterrestrial intelligence, to see if that can give us some clues about the probability of the emergence of life.

3

THE FERMI PARADOX

In April 1960, the American astronomer Frank Drake did something quite extraordinary. It was also – as far as his scientific career was concerned – highly risky. Drake turned the radio telescope at the National Radio Astronomy Observatory in West Virginia to point at two stars which resembled our Sun. Drake was attempting to listen for signs of extraterrestrial intelligence. This represented the first attempt to intercept extraterrestrial communication, and it made Frank Drake a pioneer of the search for extraterrestrial intelligence (SETI).

In 2010, Frank Drake was interviewed for a BBC documentary to discuss the search for extraterrestrial intelligence. The documentary was entitled *The Search for Life: The Drake Equation*, and this is what Frank Drake said:

> In 1960, the National Academy of Sciences asked me to convene a meeting to discuss this whole subject, and to ground it in good sound science. So I did that. I invited everyone in the world who I knew was interested in the subject to a meeting – all twelve of

them! Just twelve people who I knew were interested in extraterrestrial life.

So I thought through what it is you need to know to predict how many civilisations it might be possible to detect in our galaxy. And I realised that the number of civilisations depended on seven factors. And you can even use those factors to form an equation. So I did.

By creating his equation, Frank Drake transformed the random guesswork of predicting the existence of advanced extraterrestrial life into a theoretical framework which could be studied and analysed. The *Drake equation* can be used to calculate the number of detectable intelligent communicating civilisations, N, in our Milky Way galaxy, and is given by:

$$N = R_* \times f_p \times n_e \times f_l \times f_i \times f_c \times L$$

where:

- R_* is the number of stars formed every year.
- f_p is the fraction of those stars which have planets.
- n_e is the average number of habitable planets around those stars.
- f_l is the fraction of those habitable planets on which life emerges.
- f_i is the fraction of those lifeforms which become intelligent.
- f_c is the fraction of those lifeforms which develop into technologically-advanced civilisations capable of transmitting radio signals.
- L is the average lifetime of a communicating civilisation.

THE FERMI PARADOX

Here is Frank Drake standing in front of his eponymous equation:

While it is obviously difficult to estimate precise values for all the terms of the Drake equation, there is no obvious reason why any of the terms should be particularly small. Indeed, with more than 300 billion stars in our Milky Way galaxy it is hard to imagine why the Drake equation should produce a small value for the number of intelligent civilisations in our galaxy. In 1974, Carl Sagan used the Drake equation to predict that there were a million civilisations in our galaxy.

If that was the case, then the race was on to be the first to make contact with extraterrestrial intelligence …

6EQUJ5

In 1977, the American astronomer Jerry Ehman was sitting in his kitchen laboriously sifting through sheets of computer printouts. The data represented signals received from the Sagittarius constellation by the Ohio State University radio telescope. The telescope was also known as "Big Ear" because it was the size of three football pitches.

The Big Ear telescope formed a part of the search for extraterrestrial intelligence (SETI) pioneered by Frank Drake. The basis for the search was the idea that a technologically-advanced extraterrestrial civilisation might have broadcast a decipherable radio message in order to make themselves known to other advanced civilisations in the universe.

The printouts were of dull numeric data: sometimes a blank space, sometimes the number one, sometimes the number two. The prevalence of the ones and twos showing a low received signal strength. If you were lucky, the number three might make an appearance. And it was all repeated endlessly over countless sheets.

Most of the data was uninteresting, however, one remarkable string of six characters on the printout caught the attention of Ehman. It even looked like a word! The string of characters was "6EQUJ5". Was this a signal from an alien civilisation?

The vertical "6EQUJ5" string can be seen in the following image, a photograph of the actual printout. Ehman was so surprised at the appearance of the message that he famously wrote the word "Wow!" in the margin of the printout:

THE FERMI PARADOX

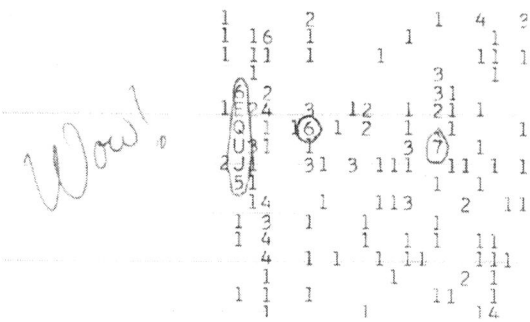

Ever since, this has come to be known as the Wow! signal.

So what does "6EQUJ5" stand for? Any Scrabble aficionado will tell you that if you get a letter Q without a letter U then you are in big trouble. So is the positioning of the U after the Q significant? If this is a message from an alien civilisation, then what are they trying to tell us? That they are good Scrabble players?

The Wow! signal became something of a cultural phenomenon. There is even a range of geeky 6EQUJ5 T-shirts available (just perform a Google image search for "6EQUJ5 T-Shirt").

Unfortunately, when you dig a bit deeper, you find all is not what it seems.

It emerges that the letters and numbers (which compose the "6EQUJ5" message) merely correspond to the strength of the signal, with the letter Z representing the highest signal strength, and the number 1 representing the lowest signal strength. This is clear in the following graph, with the coded letters and numbers on the vertical axis. You will see from the graph that when the 6EQUJ5 message is plotted it produces a very smooth curve: a normal bell-shaped distribution, a perfectly smooth "blip". All evidence of intelligent Scrabble-playing aliens has disappeared:

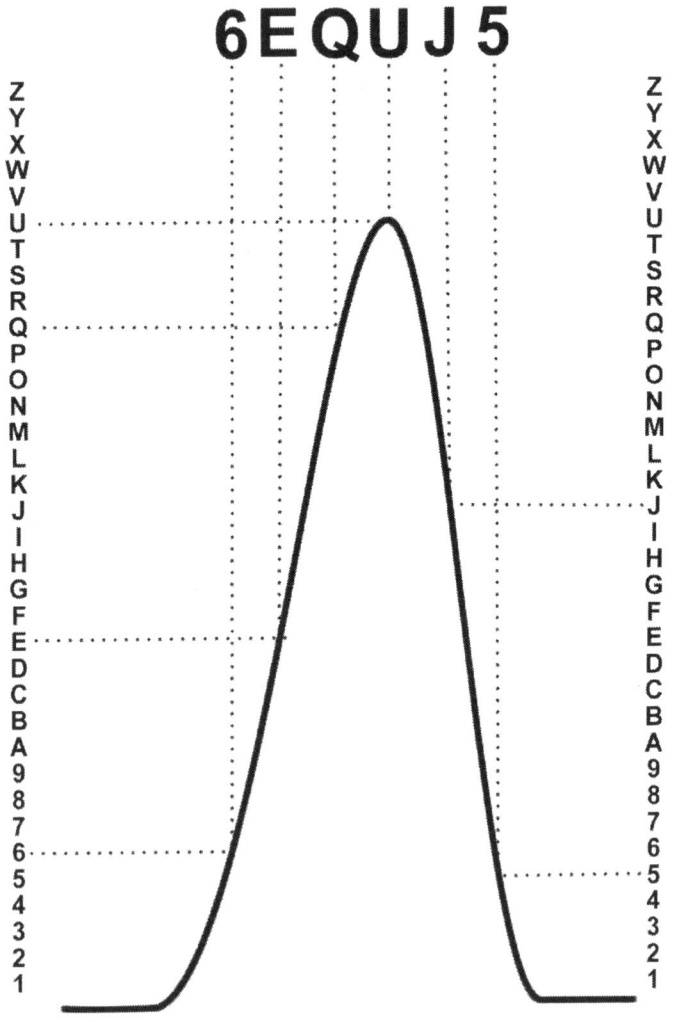

In that respect, the much-publicised 6EQUJ5 "message" appears to be rather misleading – it would have been more accurate to display the signal strength in the above graphical form rather than in alphanumeric characters. Simply put: to convert a perfectly smooth blip into a pseudo-English character string seems rather misleading.

Surely if alphanumeric characters had not been used, or if Jerry Ehman had written "This deserves further investigation" instead of "Wow!" then the signal would have gained nowhere near as much fame.

Paul Davies has written a book about extraterrestrial intelligence called *The Eerie Silence*. In that book, he examines the Wow! signal and suggests that maybe the blip is a message sent from an extraterrestrial beacon – like a flashing signal from a lighthouse. However, as Paul Davies goes on to explain, if the signal was truly a beacon then we would expect it to repeat at regular intervals – just like a lighthouse. Unfortunately, the Wow! signal has never been detected again.[6]

The Wow! signal is generally accepted to be our best candidate for a radio signal from an extraterrestrial civilisation. Given the rather misleading nature of the printout format, it is not impressive evidence. Unfortunately, the importance of being able to differentiate between accuracy and hype will be something of a theme of this chapter.

---

[6] A similar beacon signal was intercepted by the crew of the *Nostromo* commercial spacecraft in the movie *Alien*. In that case, the beacon was intended as a warning signal. So maybe we should be careful if we ever receive a genuine beacon signal …

Other attempts to detect extraterrestrial life have been based on sending robotic probes to the planets of our Solar System. Ever since H.G. Wells introduced the idea of a Martian invasion in *The War Of The Worlds*, the public has been fascinated by the possibility of life on Mars. Mars has long been regarded as the planet in our Solar System which might be most suited for life, so it was no surprise that the first robotic probe was sent to Mars as part of the Viking program of the 1970s.

Unfortunately, the Viking mission revealed the surface environment of Mars to be extremely hostile to life, with bitter cold, a thin atmosphere, and lethal ultraviolet radiation. As a result, the recent Curiosity rover – which landed on Mars in 2012 – was equipped with a drill to take soil samples from beneath the surface. Curiosity continues to send back stunning photographs from the Gale crater on Mars (as shown in the following image).

So, apart from a rather misleading "alphanumeric string" from deep space, we have so far detected absolutely no evidence of life anywhere in the universe apart from on Earth.

SETI has drawn a blank.

The Fermi paradox

Enrico Fermi was one of the giants of 20th century physics. Fermi created the world's first nuclear reactor in Chicago in 1942 and has been called the "architect of the nuclear age". If you have read my fifth book, you will know that Fermi predicted the existence of the neutrino. You will also know that the matter particles in the universe are called *fermions*, named after Enrico Fermi. When all the matter particles in the entire universe are named after you, that's when you know you've made it big.

Here is a photograph of Enrico Fermi:

In 1944, Fermi joined the Manhattan Project based in Los Alamos, New Mexico, which was developing the world's first atomic bomb. Fermi observed the Trinity bomb test in 1945, and accurately estimated the explosive power of the bomb by dropping strips of paper into the blast wave.

After the war, in 1950, Fermi returned to Los Alamos as a consultant. It was about the time that flying saucer mania was sweeping America. Legend has it that during one lunchtime in Los Alamos, during a conversation about the flying saucer craze, Fermi suddenly asked his colleagues: "Where is everybody?" Fermi's question seemed out-of-the-blue, but his colleagues laughed because everyone instantly knew what he was talking about. Fermi was wondering why we had not received clear visitations from aliens – after all, the galaxy was supposed to be teeming with alien life.

Though Fermi's observation was not treated too seriously at the time, it is not easy to find a solution. It might be imagined that the distances involved in the galaxy are too large for extraterrestrial travellers, but this is almost definitely not the case. In 1975, Michael Hart calculated that even if it takes thousands of years to travel from one star to the other, the age of our galaxy – at least ten billion years old – should ensure that life would have spread throughout the galaxy within a million years or so. This viewpoint is also expressed by Paul Davies in *The Eerie Silence* who suggests that colonisation of the galaxy would have occurred in stages, one star at a time. On that basis, the colonisation of the galaxy by one of the many supposed alien civilisations in the galaxy would appear to be inevitable, and would surely have happened by now. As Paul Davies says: "It takes only **one** such community somewhere in the galaxy to present us with Fermi's awkward conundrum."

Fermi's question has achieved considerable fame, and is now called the *Fermi paradox*.

Fermi is regarded as one of the archetypal geniuses, and we have to take his objection seriously. While it is not easy to find an answer, it does seem that many astrobiologists are now coming to something of a consensus about the likeliest solution to the Fermi paradox. To solve the Fermi paradox, it appears that we must examine the regions in our galaxy where life might possibly develop …

Exoplanets

A planet which orbits a star other than the Sun (i.e., a far distant planet outside of the Solar System) is called an *exoplanet*. As exoplanets are small, dark, and very distant, they are very difficult to detect. Because they orbit a star, the light from the star will drown out any light emitted or reflected from the exoplanet. Hence, exoplanets are not usually observed directly, but are instead detected via indirect methods.

One indirect method for detecting exoplanets is based on the principle that an orbiting exoplanet will very slightly pull the star from its central position. This slight deviation in the star's motion can be detected by considering modifications in the wavelength of the light emitted by the star due to the Doppler shift. This is the same principle which causes a siren on a car to sound higher in pitch when the car is approaching, and lower in pitch when the car is receding. Extremely small variations in the star's velocity – as little as one metre/second – can be detected by this method, which is called the *radial velocity* method.

Exoplanets are certainly not rare in the universe, and several thousand have already been detected. The exoplanets which have been detected so far have been classified into several groups. The first group to be detected were the *Hot Jupiters*, which are extremely large gas giants – like the planet Jupiter. However, unlike Jupiter, these exoplanets orbit close to central stars (and therefore have very high surface temperatures). These exoplanets travel at high orbital speeds, maybe only taking a few days to complete an orbit. The large masses of the Hot Jupiters, in combination with their tight fast orbits, explains why they were the first exoplanets to be discovered in the mid-1990s using the radial velocity method (high mass and speed causing the central star to wobble).

NASA's Kepler Space Telescope was launched in 2009 with the specific task of detecting Earth-sized exoplanets. The detection method used by the Kepler mission was different from the radial velocity method. Instead, the detection method was based on the principle that the brightness of a star will dip slightly when an exoplanet passes in front of it. This is called the *transit* method for detecting exoplanets. According to NASA, Kepler "launched the modern era of planet hunting".

The Kepler telescope is trained on a narrow region of space, roughly the size of your fist held at arm's length. In that region of space, Kepler detects the amount of light emitted from approximately 150,000 stars. Any repeated temporary reduction in the brightness of one of those stars is interpreted as the transit of an exoplanet passing in front of the star.

From examining the Kepler data, it is clear that exoplanets are far from rare in our galaxy. In fact, it would appear that there is the staggering number of between 100 billion and 400 billion exoplanets in our Milky Way galaxy. That means that, on average, there is one exoplanet for each star in the sky. So when you look up into the night sky, remember that each of those stars is likely to have a planet orbiting it.

The transit method used by Kepler is able to detect smaller exoplanets than the older radial velocity method, and is able to detect Earth-sized exoplanets. The Kepler mission was able to detect another grouping of exoplanets called the *super-Earths*. The super-Earths are *terrestrial* planets (which means they are rocky like the Earth, as opposed to gaseous) but can have masses up to ten times larger than the Earth.

Exoplanets have tremendously variable characteristics. In his book *The Copernican Complex*, Caleb Scharf considers the varied properties of the super-Earth exoplanets:

> *Many appear to have huge atmospheres, possibly containing lots of hydrogen. Some of these bulky objects are likely covered in vast quantities of water. They may be frozen solid. They may also be awash in a global ocean that reaches to almost unimaginable depths – tens, even hundreds of miles – with pressures and temperatures such that the physical and chemical behavior of water becomes unlike anything we experience on Earth. Others may have a modest splash of water, or none at all. But many should be persistently volcanic.*

A theme we get from these exoplanets is that there is a tremendous diversity. Exoplanets have a huge range of different sizes, orbits, chemical composition, and surface conditions. Some exoplanets have rings of dust or ice, like Saturn, while others have moons.

NASA has released a range of imaginative travel posters from the fictional "Exoplanet Travel Bureau" which advertise travel to a variety of known (genuine) exoplanets. The posters emphasise the various attractive characteristics of each exoplanet, hopefully persuading you to take a holiday there. As an example, one of the advertised exoplanets is Kepler-16b, and its featured characteristic is that it orbits a pair of stars, like Luke Skywalker's planet Tatooine in *Star Wars*. As NASA say, "the movie's iconic double-sunset is anything **but** science fiction":

The habitable zone

There may be billions of exoplanets in our galaxy, but what we are really interested in is how many of those exoplanets possess the right conditions for life. A key mission objective of Kepler is to determine how many exoplanets have the properties which are believed to be essential for life to develop.

What properties are considered essential for life? Well, in 2009 the highly-popular BBC TV show *Top Gear* sent its three presenters on a road-trip across Bolivia. At one point, Jeremy Clarkson, Richard Hammond, and James May found themselves driving along the only road across the Atacama Desert. The Atacama Desert is quite beautiful. You could have a perfectly pleasant day out in the Atacama Desert. The scenery is stunning, and the temperature is not overwhelming. You could take a deckchair and a book and enjoy yourself. However, if you did that, you would be the only form of life of any kind for many miles in any direction. As Clarkson said: "We were in the Atacama Desert, where there is no life at all – not even on a cellular level. Richard Hammond was the smallest living organism for miles."

There is absolutely no life at all in the Atacama Desert: no plants, no insects, no bacteria. Absolutely nothing. And yet you could enjoy a very pleasant day sitting in the desert. How can that be the case? It is because the Atacama Desert lacks just one property which is absolutely essential for life: no rain ever falls on the Atacama Desert. It has no water.

As water appears to be essential for life, one of the most important factors in determining if an exoplanet could support life is the climate of the exoplanet. Is it too hot or too cold? A planet which is either too hot or too cold to support liquid water would appear unable to support life.

The temperature of a planet is dependent on the orbital distance of the planet from the central warming star. In our own Solar System, the planets Mercury and Venus orbit too near the Sun to support liquid water: water could only exist as vapour, whereas Mars is now too cold for liquid water. The Earth lies within those two orbits, with surface temperatures allowing the continual presence of liquid water. There is a similar orbital range around all stars within which liquid water could exist on a planet. This band is called the *habitable zone*.

The following diagram shows how the Earth is the only planet in the Solar System to lie in the narrow habitable zone band:

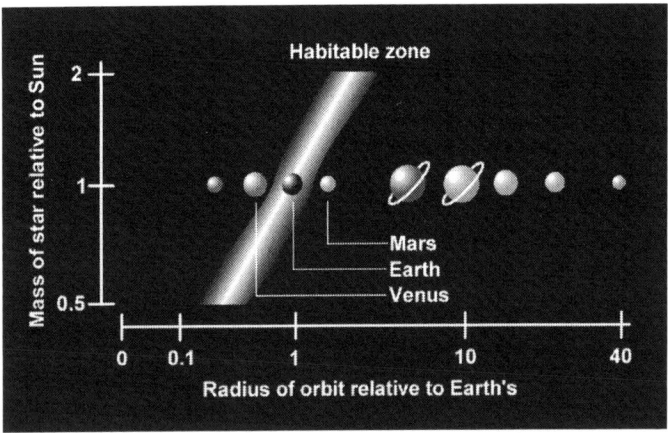

You will see that the habitable zone band is at an angle, showing that a smaller central star would produce less heat, and so the habitable zone would lie closer to the star.

At the moment, it seems almost every month there is the announcement of the discovery of an "Earth-like" habitable exoplanet which could potentially support life. But how accurate are these claims?

THE FERMI PARADOX

As an example, during the writing of this book, there was a major announcement of a discovery by the European Southern Observatory (ESO) telescope in the dry Atacama Desert (because lack of water vapour in the air and no clouds allows great astronomy). The discovery was of an exoplanet with a mass between 1.3 and three times as much as Earth, found in the habitable zone of our nearest star, Proxima Centauri. The exoplanet was given the name Proxima Centauri b (also known as Proxima b).

Proxima b orbits its central star quite closely, at only 5% of the distance between the Earth and the Sun. However, the central star – Proxima Centauri – is a type of star called a red dwarf, which is a smaller, dimmer, and cooler type of star than the Sun. Hence, even though Proxima b has a close orbit, it still lies within the habitable zone of the star.

The ESO announcement came complete with an artist's impression of the surface of Proxima b. Coming over the horizon in the distance you will see the red dwarf star, Proxima Centauri, around which the exoplanet is orbiting:

The announcement generated enormous media coverage. As an example, the front page story of *New Scientist* magazine (27th August 2016) proclaimed "We've found an Earth-like planet around our nearest star". The headline in the *Times* newspaper was "New planet could be our home away from home".

So much media excitement has been generated that there is now a plan to send an unmanned spacecraft to Alpha Centauri to investigate (Alpha Centauri is the three-star system which includes the star Proxima Centauri around which the exoplanet Proxima b orbits). But, in order to travel four light years, this can be no ordinary spacecraft.

The Russian billionaire Yuri Milner is funding a project called *Breakthrough Starshot*, the aim of which is to send a tiny laser-powered spacecraft to Alpha Centauri. Actually, the spacecraft is so small – a lightweight electronic wafer not much larger than a postage stamp – that it is called a *nanocraft*.

The nanocraft will be attached to a three-metre sail which will provide the propulsion. But the sail will not be powered by wind – it will be powered by light. As described in my third book, although light has no mass, it has momentum. Therefore, light beams can be used to exert pressure onto an object, so light shone onto the sail can provide the motive power for the nanocraft. However, the amount of force applied by a beam of light is extremely small, which explains why the nanocraft must be so lightweight.

The following image shows an artist's impression of the light sail and the small nanocraft to be used in Breakthrough Starshot. The sail is intended to be three metres across, which gives you an impression of the small size of the nanocraft (shown in the middle):

THE FERMI PARADOX

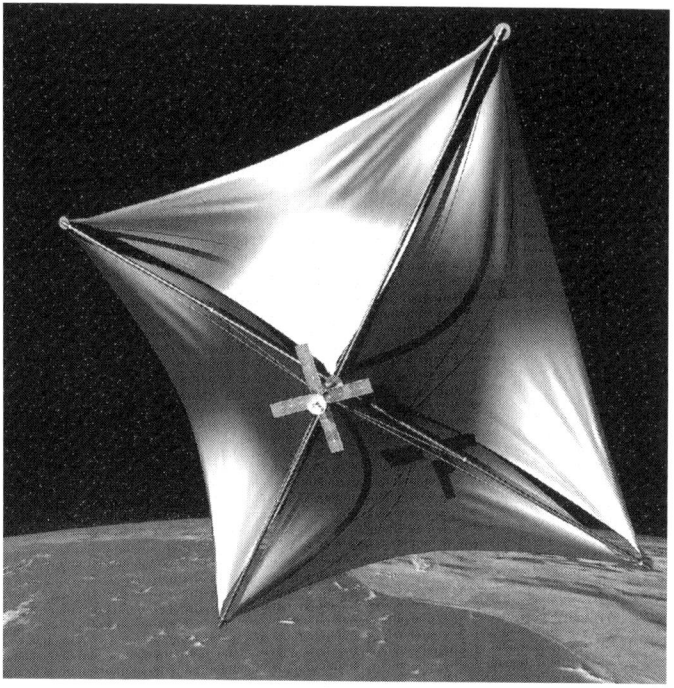

The aim is to power the craft by shining powerful Earth-bound lasers at the sail. The intended power of the lasers would accelerate the spacecraft to a fifth of the speed of light in just an hour, by which time, the nanocraft would have reached Mars. After that initial "kick", the nanocraft would reach the Alpha Centauri system in less than twenty years.

If all of this sounds crazily ambitious, that's because it **is** crazily ambitious. The required total laser output power would be 100 gigawatts, which is equal to the total electrical power consumption of France. It would be extremely difficult to focus such a powerful beam of light through the Earth's turbulent atmosphere. Also, travelling at a hundred million miles an hour (yes, it really would be travelling that fast!), even a fleck of interstellar dust would be enough to destroy the nanocraft.

So is the plan to send a nanocraft to investigate Proxima b overly-ambitious? The project is clearly going to generate some eye-catching headlines: "Spacecraft to travel at near speed-of-light to nearest Earth-like planet". But does this accurately reflect the reality of the scientific situation, or is it merely designed to generate publicity?

With this thought in mind, let us return to consider the "Earth-like" exoplanet, Proxima b. The artist's impression of the surface (shown earlier) is certainly beautiful, but is it accurate? When you start to dig deeper, you find things are not all they seem.

The artist's impression shows a rocky surface, but we know nothing about the actual surface of Proxima b. The exoplanet was detected indirectly via the radial velocity method (deflection of the central star) which tells us nothing about the chemical composition of the exoplanet. This assumption of a rocky surface is purely based on the observation of planets of a similar mass in our Solar System. But without direct observation of the exoplanet surface, we cannot be sure what it is like.

Because the exoplanet is so relatively close, direct observation might be possible using a space telescope. This would pin down the mass, size, and density of the planet. Because of the great distance, and the corresponding high resolution which would be required, this would only be possible using the large James Webb Space Telescope, which is NASA's replacement for Hubble and is due to launch in 2018. Even then, the exoplanet is so small and distant it would only appear as a single pixel of light in the telescope. Hence, in the absence of direct imaging, we do not know if Proxima b has an atmosphere, or if it has any water.

So, clearly, calling Proxima b an "Earth-like planet" in sensationalist newspaper headlines seems premature. This is certainly the case when we consider additional data which makes Proxima b seem definitely unlike the Earth.

Proxima b, like so many exoplanets, orbits so close to its central star that it is constantly bombarded with deadly ultraviolet and X-ray radiation from the star. The intensity of the X-ray radiation from a red dwarf is 400 times greater than that experienced by the Earth.

Also, because Proxima b is so close to its central star it is likely to be tidally-locked to the star, which means that gravity pulls the planet so tightly that the same side of the planet always faces the star – just as the same side of the Moon always faces the Earth. Hence, one side of the planet would be permanently baked by the heat from the star, while the other side of the planet freezes in the darkness of space.

There is a fictional (and scientifically accurate) example of the difficulty in finding an exoplanet which is suitable for life. In the movie *Interstellar*, astronauts leave Earth in order to colonize a new planet to ensure the survival of humanity. However, the first two exoplanets they inspect are certainly not suitable, in fact, they are completely hostile to life. The first exoplanet turns out to be covered with a ocean which experiences kilometre-high tsunamis, while the second exoplanet turns out to be completely frozen with an atmosphere of toxic ammonia. It appears that this is a more realistic picture of what we might actually expect to find on these exoplanets.

So, once we cut through the hype, this discussion appears to raise an important question …

Is the Earth special?

Peter Ward is Professor of Geological Sciences and Donald Brownlee is Professor of Astronomy, both at the University of Washington in Seattle. They are acknowledged experts in their fields. In the year 2000, Ward and Brownlee published a book called *Rare Earth* which, it has been said, "hit the world of astrobiologists like a killer asteroid".

In *Rare Earth*, Ward and Brownlee listed many other additional requirements for any habitable planet, leading them to the conclusion that the Earth might well be the only planet suitable for advanced life in the galaxy. Ward and Brownlee called this the *Rare Earth Hypothesis*:

> *Ever since Polish astronomer Nicholas Copernicus plucked it from the center of the universe and put it in orbit around the Sun, Earth has been periodically trivialized. We have gone from the center of the universe to a small planet orbiting a small, undistinguished star in an unremarkable region of the Milky Way galaxy – a view now formalized by the so-called Principle of Mediocrity, which holds that we are not the one planet with life but one of many. Various estimates for the number of other intelligent civilizations range from none to ten trillion.*
>
> *If it is found to be correct, however, the Rare Earth Hypothesis will reverse that decentering trend. What if the Earth, with its cargo of advanced animals, is virtually unique in this quadrant of the galaxy – the most diverse planet, say, in the nearest 10,000 light years? What if it is utterly unique: the only planet with animals in this galaxy or even in the visible universe?*

THE FERMI PARADOX

In one of their proposals, Ward and Brownlee extended the idea of the habitable zone of a planet orbiting a star to suggest that there is also a habitable zone in each galaxy, a certain distance from the galactic centre. The region near a galactic centre is star-packed, and the constant upheaval from stellar collisions and supernovae explosions might not allow the billions of years of relative calm for advanced life to develop. At the other extreme, the outer regions of galaxies have fewer stars and would lack the heavy elements used to build rocky planets (because heavy elements are generated by fusion in the interior of stars). Our Solar System happens to be located at just the right distance from the centre of our Milky Way galaxy to allow life to develop:

It is clear that combining the two habitable zone requirements — around a star and also around the galactic centre — places severe restrictions on the number of habitable planets in a galaxy.

Ward and Brownlee also stress the importance of a stable environment for the development of life. It is known that the Sun is becoming brighter with time, and it is now 30% brighter than it was when the Earth was formed. About four billion years from now, the Sun will expand rapidly to become a red giant, its brightness increasing over five thousand times. At that point, the radius of the Sun will have expanded until it reaches the orbit of the Earth – extinguishing all life on the planet.

However, all that bad news lies in the far distant future. The Sun has been an unusually stable – and safe – source of energy for the Earth for approximately four billion years. It is known that biological evolution requires vast periods of time: probably billions of years. Hence, the stability of the Sun has allowed plenty of time for life to develop on Earth. However, stars larger than the Sun become brighter much faster. A star 50% more massive than the Sun would have become a red giant too quickly for life to evolve on any orbiting planet.

The stability – and safety – of the Earth over billions of years has also been ensured by the presence of Jupiter and Saturn orbiting outside the Earth. This is because it is believed that another requirement for the development of intelligent life is that there is a necessity for a planet to be protected from continual violent collisions with comets or asteroids. Stephen Hawking has said that a collision with a comet or asteroid greater than twenty kilometres in diameter would result in the mass-extinction of complex life. It is now believed that the gravity of massive Jupiter and Saturn located outside the orbit of the Earth acts to catch many asteroids and comets entering the Solar System before they can collide with the Earth. As an example, the comet Shoemaker-Levy 9 was observed colliding with Jupiter in 1994, leaving a scar on its surface.

Ward and Brownlee also emphasise the importance of *plate tectonics* in creating a habitable Earth. Plate tectonics is the movement of the crust of the Earth. This creates mountain ranges (where two continental plates crush together) and volcanoes. Volcanic activity appears to be particularly important because of the volume of carbon dioxide which is released into the atmosphere by volcanoes. Carbon dioxide is a greenhouse gas which captures outgoing infrared radiation and raises the temperature of the surface of the Earth. It is believed that the release of greenhouse gas was responsible for saving the Earth from previous ice ages which would have resulted in the irreversible freezing of the entire Earth. It is the temperature-mediating properties of greenhouse gases which has allowed liquid water to reside on the surface of the Earth for four billion years. As Ward and Brownlee say in *Rare Earth*: "Greenhouse gases are keys to the presence of fresh water on this planet and thus are keys to the presence of animal life." We would perhaps be wise to remember that if it wasn't for greenhouse gases the Earth would not be habitable at all. Global warming is not always a bad thing!

While other planets – such as Mars – have high, isolated volcanoes, only the Earth has long mountain ranges at the edge of continental plates. For this reason, it is believed that Earth is the only planet in the Solar System with tectonic plates, and plate tectonics might be extremely rare in the universe as a whole.

The metallic core of the Earth also plays an important role by generating a magnetic field which protects the Earth from lethal solar radiation. The optimistic analysis of exoplanet habitability does not take into account whether or not the exoplanet has a similar protective field. The situation is summed-up by Prof. Don Pollacco from the University of Warwick who is an expert on exoplanets: "As we learn things about what makes the Earth habitable, things like the magnetic field become really important. We can't measure the magnetic field of an exoplanet, so we just forget about it."[7]

Caleb Scharf is the director of Columbia University's Astrobiology Center. In his recent book *The Copernicus Complex*, Scharf starts by reminding us of the Copernican revolution in science, where the picture of the Earth as the centre of the universe was replaced by a model in which the Earth is nothing special: an unremarkable planet orbiting an unremarkable star. However, as Scharf realises: "While we cannot be at the center of what we now know to be a centerless universe, we appear to occupy a very interesting place within it – in time, space, and scale."

As Scharf suggests, there does appear to be something special about the Earth in terms of its position and its size. With this in mind, whenever I read about the habitable zone for exoplanets – the thin band of space in which life is supposedly possible – I am always reminded of the mathematical construction known as the *Mandelbrot set* (considered in my fourth book). The Mandelbrot set is constructed from a very simple iterative formula, producing a value which is defined for every point in the two-dimensional plane. Hence, it is possible to generate a rectangular image representing the Mandelbrot set, and this turns out to be a very pretty image which has been used in many posters and computer backgrounds. However, perhaps what is not widely realised is that the only truly pretty part of the Mandelbrot set is the very thin, crinkly boundary of the set:

---

[7] *Where should we look for alien life?*, BBC website, http://tinyurl.com/wherealienlife

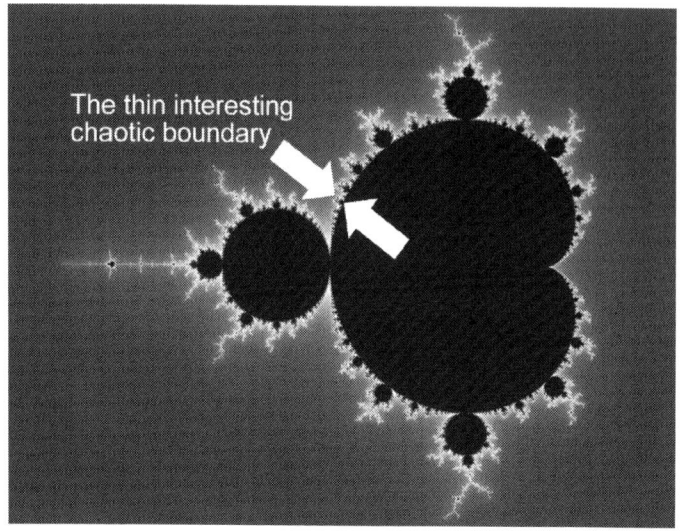

To my mind, this thin, interesting, chaotic boundary has obvious parallels with the habitable zone around stars (and galaxies): it is where everything interesting happens. And, what is most interesting from the point of view of the discussion in this book, the interesting boundary is extremely thin. The vast, empty regions of the image are not interesting. Does the Earth lie on the equivalent of that extremely thin Mandelbrot boundary? Is the Earth special in that regard?

There is a famous quote from Stephen Hawking in his book *A Brief History of Time*: "We have developed from the geocentric cosmologies of Ptolemy and his forebears, through the heliocentric cosmology of Copernicus and Galileo, to the modern picture in which the earth is a medium-sized planet orbiting around an average star in the outer suburbs of an ordinary spiral galaxy, which is itself only one of about a million million galaxies in the observable universe." In that quote, Stephen Hawking is suggesting the Earth is nothing special – completely mediocre. And, to the extent that he describes, he is obviously correct. However, with all due respect to Stephen Hawking, I believe in this case he misses the point. It is now being realised that it is this very mediocrity that makes the Earth special. The Earth is completely and perfectly middling: it lies in the narrow habitable zone. The Earth is not the biggest, it's not the smallest. It's not the hottest, it's not the coldest. It lies in a relatively safe planetary system where nothing very exciting ever happens, no planetary collisions or extinction-level asteroid strikes. At first glance, nothing about the Earth appears to be remarkable – apart from the fact that it is the only place in the universe that we know of where life exists. Ironically, it is the perfect mediocrity of the Earth which makes it special.

As Caleb Scharf says in *The Copernicus Complex*:

> Life is a collection of phenomena at the boundary between order and chaos. Too far away from such borders, in either direction, and the balance for life tips toward a hostile state. Life like us requires the right mix of ingredients, of calm and chaos – the right yin and yang. There are obvious parallels to the concept of a habitable zone, which proposes that a temperate cosmic environment for a planet around a star exists within a narrow range of parameters.

The Copernican principle quite correctly proposed that the Earth orbited the Sun – just like any other planet. But this message of the Copernican principle – that the Earth is mediocre, "just another planet" – has then been incorrectly applied to infer that all other planets must be like the Earth. Hence, the Copernican principle has been used to predict an abundance of life on all exoplanets in the universe. However, we are now moving beyond the domain of application of the Copernican principle. As an example, chaos theory was only discovered in the nineteenth century, exoplanets were only discovered in the twentieth century, so we should no longer be relying on a sixteenth century principle (the Copernican principle) to determine the probability of extraterrestrial life. Science does advance.

Caleb Scharf believes our place in the universe is special, and the raw Copernican Principle is no longer sophisticated enough to capture the whole truth:

> The Copernican Principle may have reached the end of its usefulness as an all-encompassing guide to certain scientific questions. The Copernican Principle is both right and wrong, and it's time we acknowledge that fact.

So we now see we have discovered perhaps the most likely solution to the Fermi Paradox. Earlier in this chapter it was described how it might seem likely that every term in the Drake equation has a large value, in which case the predicted number of advanced civilisations in our galaxy would be huge: Carl Sagan used the Drake equation to predict a million civilisations in our galaxy. However, there is of course a different way of interpreting the Drake equation, as explained in the *Rare Earth* book:

> Any factor in the equation that is close to zero yields a near-zero final answer, because all the factors are multiplied together.

In other words, it only takes one of the factors in the Drake equation to be small in order to send the number of predicted civilisations to come crashing down. And, from our discussion in this chapter, it would appear it is the n_e term in the Drake equation which is small (the n_e term is the average number of **habitable** planets orbiting a star) and that is the reason why we do not observe a universe teeming with life.

As Matthew Cobb says in the recently-released book *Aliens* (edited by Jim Al-Khalili), our current knowledge "leads us to the conclusion that the answer to the Fermi paradox is that its starting point is probably wrong. There are no alien civilisations."

But, even if all the interesting processes (life) occur within a narrow, chaotic band, we still need that vast uninteresting expanse of the rest of the universe in order for the Earth to be so interesting. It is not the Earth that is mediocre – it is the rest of the universe that is mediocre. The Earth is, quite frankly, an amazing place. As Ward and Brownlee say in *Rare Earth*: "Earth seems to be quite a gem".

Are we alone in the universe? It appears that might well be the case.

I am reminded of a humorous quote by the great British comedian, Peter Cook: "As I looked out into the night sky, at all those infinite stars, it made me realise how insignificant … they really are."

Implications for fine-tuning

Finally in this chapter, let us return to the question of whether or not the universe is fine-tuned for life. How has this discussion of extraterrestrial life affected the likelihood of that hypothesis being true?

If you remember, at the end of the Bayesian analysis in the previous chapter, it was suggested that it we want to calculate whether or not the universe is fine-tuned, we only need to get a definitive answer to the "probability of life appearing in general". Remember, if life emerges easily in all universes, then it has no need of additional helpful fine-tuning. Conversely, if life does not emerge easily, then life is delicate and rare and it needs all the help it can get from fine-tuning in order to emerge.

Because of the convincing evidence of the Fermi paradox – and its likely resolution in the form of the Rare Earth Hypothesis which suggests that life requires very particular and unusual circumstances in order to emerge – I am going to suggest that the "probability of life appearing in general" should be set to a low value, certainly a lower value than has been suggested in the past by astrophysicists such as Carl Sagan. This might appear to be a controversial conclusion, but it seems to be in line with the recent conclusions of many astrobiologists.

And, if you remember back to the Bayesian fine-tuning equation presented in the previous chapter, a low value for the "probability of life appearing in general" lends support to the fine-tuning argument. As Caleb Scharf says: "There are signs that we inhabit a somewhat unusual place, and there is a hint of an expansion to the notion of cosmic fine-tuning."

The apparent fine-tuning problems are many and varied. Some may not be true problems at all: it is possible that life would still be possible even with considerable variation in the values of the parameters. However, while some parameter limits might be loosened, it appears that many limits still remain. From our discussion of the Fermi paradox, it appears that even a failure to produce liquid water would prohibit the emergence of life. It has also been suggested that life might be possible in a completely different form if the physics of the universe was completely different. One example is the possibility that silicon-based life might emerge in a universe which is unable to produce carbon.

However, life based on alternative chemistry does nothing to avoid the fundamental fine-tuning problems. Considering silicon-based life, for example, silicon is just another heavy element (like carbon) which can only be produced in the interior of stars by fusion reactions. The fine-tuning coincidences on which heavy-element production depends are just as relevant for silicon as they are for carbon. As John Barrow says in his book *The Constants of Nature*: "The argument is not really changed if beings are possibly based upon silicon chemistry or physics. All the elements heavier than the chemically inert gases of hydrogen, deuterium and helium are made in the stars like carbon and require billions of years to create and distribute."

Another criticism of silicon-based life is presented by Paul Davies in *The Eerie Silence*: "There has been some speculation that silicon could substitute for carbon, a conjecture that got as far as an episode of *Star Trek*, but it hasn't been pursued very seriously by biochemists because silicon can't form the extraordinary range of complex molecules that carbon can."

As another example, the chemistry of a simplistic alternative universe might only allow the existence of hydrogen and helium. However, as Caleb Scharf says: "It does seem hard to imagine how a universe of just hydrogen and helium could give rise to structures with the complexity seen in carbon-based life."

THE FERMI PARADOX

So it appears that there is a real problem here, and we need to find a solution.

I believe that, in order to solve the many fine-tuning problems, we need to avoid approaches such as the strong anthropic principle, which promise simple and easy solutions to difficult fine-tuning questions. There is no substitute for good physics. We will have to rely on our talent, ingenuity, and hard work to solve these problems.

On that basis, I can assure you that the remainder of this book will be based on conventional theoretical physics. There will be no multiverses, no anthropic reasoning, and no *Star Trek* silicon-based lifeforms in sight.

And, once we start using conventional theoretical physics to analyse these problems, we find another possible alternative to fine-tuning emerges. Maybe the laws of physics are unique, and completely determine the values of the fundamental constants to be suitable for the emergence of life? Paul Davies makes this point in his book *The Goldilocks Enigma*: "We have no idea whether the various parameters of interest are actually free and independent, or whether they will turn out to be linked by a more comprehensive theory, **or possibly even determined completely by such a theory.**" This is in line with Einstein's famous quote: "What I am really interested in is whether God could have made the world in a different way; that is, whether the necessity of logical simplicity leaves any freedom at all."

If it is the case that the laws of physics, and the values of the fundamental constants, could take no other conceivable form then the universe would not be fine-tuned because **no fine-tuning of the values would be possible.** In that case, we would have solved the fine-tuning problem.

Might it be the case that the laws of physics are uniquely determined to have values which are suitable for life? Is life inevitable?

We are going to have to consider some theoretical physics...

4

THE COSMOLOGICAL CONSTANT

In the first two decades of the 20th century, our view of the universe was a lot smaller than it is today. The existence of other galaxies was not certain, so all the stars were believed to lie within our own Milky Way galaxy. What is more, the position of the stars appeared to be fairly stable. There was certainly nothing to suggest that the universe might be rapidly expanding.

So when Einstein published the theory of general relativity in 1915, he was left with something of a conundrum. The theory suggested that a stable universe was not possible as gravity would eventually pull all the matter together. How could Einstein resolve this conflict between observations and the prediction of his theory? Unusually for Einstein, he rejected the prediction of his elegant and revolutionary theory, and tried to "fudge" the result in order to produce a stable universe.

In order to understand what Einstein suggested, let us consider the equation for general relativity (which was derived in my previous book):

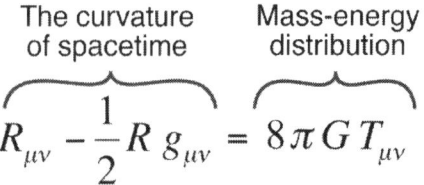

$$R_{\mu\nu} - \frac{1}{2} R\, g_{\mu\nu} = 8\pi G\, T_{\mu\nu}$$

You will see that the equation actually predicts something quite simple. It states that the curvature of spacetime (on the left-hand side of the equation) is dependent on the distribution of mass and energy. It is perhaps best described by a famous quote by John Wheeler: "Spacetime tells matter how to move; matter tells spacetime how to curve."

To preserve a stable universe, Einstein added a factor, Λ, to his equation which was effectively a form of repulsive gravity. This acted to move the stars apart to balance the inward contraction. Einstein called this additional term the *cosmological constant*:

$$R_{\mu\nu} - \frac{1}{2} R\, g_{\mu\nu} + \Lambda\, g_{\mu\nu} = 8\pi G\, T_{\mu\nu}$$

⬆ Cosmological constant

It is interesting that Einstein placed his correction factor on the left-hand side of the equation, thus modifying the curvature of spacetime. He might have chosen to place his correction on the other side of the equation, thus modifying the predicted energy of space. As we shall see later, this latter approach is the one favored in modern cosmology.

The priest

Georges Lemaître was a Belgian physicist who is now regarded as one of the most important physicists and cosmologists of the 20th century. Surprisingly, however, Lemaître was not a physicist by profession. Instead, in 1923, he was ordained as a Roman Catholic priest and remained in the priesthood for his entire life.

Lemaître always managed to keep a distance between his personal beliefs and his scientific research, even once arguing with the Pope over the Pope's interpretation of Lemaître's own Big Bang theory. That disagreement did not stop Lemaître from eventually becoming the president of the Pontifical Academy of Sciences.

In 1924, Lemaître realised that Einstein's cosmological constant would not be sufficient to produce a stable universe. Like a pencil balancing on a point, the apparently stable universe would actually be unstable. Even the slightest variation between the amount of matter in the universe and the value of the cosmological constant would result in a universe which either contracted to a point or expanded forever. A pencil cannot stay balanced on its point forever.

Lemaître's 1924 paper also predicted that if the universe really was expanding then the expansion could be measured by astronomical observations. Light from receding galaxies would be shifted towards the red end (longer wavelength) of their spectrum because of the Doppler effect. Lemaître realised that general relativity predicted a linear relationship between the distance of faraway galaxies and their redshift.

In 1927, the American astronomer Edwin Hubble decided to check this predicted relationship between distance and velocity. Hubble calculated the distance of galaxies by measuring their brightness, and relied on his assistant – Milton Humason – to measure the spectra of the galaxies.

By January 1929, Hubble and Humason had distance and spectra data for twenty-four distant galaxies. Hubble plotted the data on a graph, with distance on the x axis and velocity (calculated from the redshifts) on the y axis. The scatter of points lay close to a straight line, indicating a linear relationship between distance and recession velocity – just as Lemaître had predicted.

The following image shows Hubble's actual graph from his 1929 paper (which never credited the work of Humason) in the Proceedings of the National Academy of Sciences:

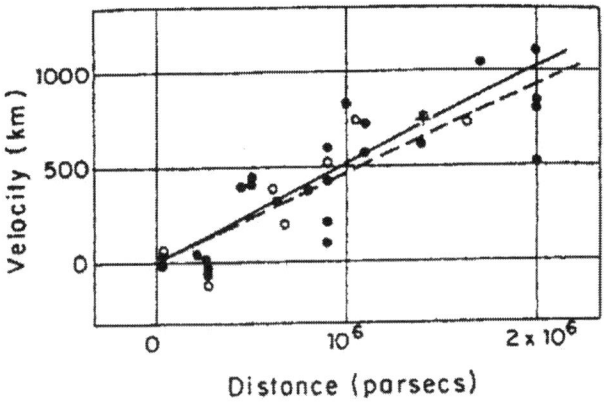

When he read about Hubble's discovery, Einstein realised his blunder in horror. An expanding universe had no need of a cosmological constant to keep it static. Einstein realised he could have predicted the expanding universe from general relativity before Hubble discovered it from observation. Einstein was later to call his introduction of the cosmological constant the "biggest blunder" of his life.

Einstein had lost the initiative and the leadership. From now on, for the rest of his career, Einstein was always playing catch-up in physics.

The return of the cosmological constant

With Hubble's discovery, it appeared to be the end of the role of the cosmological constant which was duly removed from the general relativity equation. However, in 1998, a surprising astronomical observation brought the cosmological constant back to centre stage once again.

After the initial explosion of the Big Bang, general relativity predicted that the expansion of the universe would slow down due to the gravitational attraction between the masses in the universe. The question to be answered was at what rate was the expansion slowing?

In order to answer this question, two international teams of researchers attempted to measure the rate of expansion of the universe. The teams measured the brightness of supernova explosions, which tended to all be of similar brightness and were therefore termed *standard candles*. This provided a measure of the distance to the supernova. The speed at which the supernova was receding could be calculated from its redshift.

By combining these two results, both teams discovered something remarkable: the expansion of the universe was accelerating, not slowing-down as predicted. This came as a complete surprise, and the cause of the expansion was unknown. However, it was realised that resurrecting Einstein's cosmological constant provided a possible solution. Remember, Einstein introduced the cosmological constant into his equation to provide a form of repulsive gravity needed to counterbalance the contraction of the universe. It could be this repulsive gravity which was now powering the accelerating expansion of the universe.

What was more, there was an obvious candidate for the energy required to power the expansion. In the next chapter on quantum field theory we will be examining how particles (together with their equivalent antiparticles) can be produced if enough energy is packed into a sufficiently small volume. Even in empty space, the uncertainty principle can introduce the possibility that a particle pair might be produced for an extremely short time, before annihilating each other. As a result, even in empty space, in which all particles have been removed, there will be a seething froth of these so-called *virtual particles*. There will be a corresponding energy due to the presence of the virtual particles called the *vacuum energy*, or the *zero-point energy*.

The suggestion was made that it might be this vacuum energy which was driving the accelerating expansion of the universe, with the vacuum energy playing the role of the cosmological constant in Einstein's equation. This mysterious energy was given the name *dark energy*. However, a problem arises when we try to calculate the expected value of the vacuum energy by adding all the quantum contributions from the virtual particles: we find we get an infinite result. Why is this the case? Well, quantum mechanics tells us that every particle can also be considered as being a wave (*wave-particle duality*), with shorter wavelengths representing particles with higher energy. But when we consider the vacuum energy of a volume of space, there is nothing to stop us including the contributions of particles with ever-smaller wavelengths, which represent ever-increasing energies – potentially up to infinity. The only way we can get a meaningful finite result is to have a cut-off wavelength, and not consider the contributions from any wavelengths which are shorter than the cut-off.

The logical cut-off wavelength is provided by the *Planck length*, an extremely small distance (about 10^{-20} times the diameter of a proton) which represents the shortest measurable length. Unfortunately, when we use the Planck length as the cut-off wavelength, the value of the vacuum energy is calculated to be the huge value of about 10^{93} grams per cubic centimetre. In his book *The Goldilocks Enigma*, Paul Davies calculates that this is "implying that a thimbleful of empty space should contain a million trillion trillion trillion trillion trillion trillion tonnes!" This is larger than the necessary value for dark energy by a factor of 10^{120}. According to Stephen Hawking, this outrageous value represents the "biggest failure of physical theory in history".

As a result, the only way that vacuum energy can be the cause of the accelerating expansion of the universe is if there is some cancellation effect from different types of virtual particles, greatly reducing the estimate to the observed small value. But this would represent fine-tuning to a truly incredible degree.

What is more, if the extremely small value of the cosmological constant had a much larger value then it would have stopped galaxies and stars from forming, and life would have been unable to evolve. This has driven some physicists to consider desperate measures. According to Leonard Susskind: "The notorious cosmological constant is not quite zero, as it was thought to be. This is a cataclysm and the only way that we know how to make any sense of it is through the reviled and despised anthropic principle."[8]

But surely we can do better?

---

[8] http://tinyurl.com/susskindlandscape

Naturalness

In order to understand the fine-tuning problem more clearly, it is useful to consider the principle of *naturalness* (considered in depth in my previous book).

In physics, there are some fundamental physical constants which seem to play particularly important roles in describing the universe. Most of these constants have associated units, for example, the speed of light is 3×10^8 metres/second (with the units in that case being metres per second). Note that we could change the numeric value of the constant merely by changing the units. For example, it is also correct to say that the speed of light is one foot per nanosecond. Or, equivalently, 328,000 football pitch lengths per second. Or maybe 46 Great Walls of China per second. Or even 1.6×10^8 giraffes' necks per second.

So it is clear that when we are dealing with constants that have associated units, the actual numeric value of the constant is completely dependent on the units we choose.

This is not the case for a particular group of constants which do not have associated units. These are called the *dimensionless* constants. As an example, later in this book we will be considering the fine structure constant which has a value of approximately $1/137$ – with no associated units. Because the dimensionless constants have no units, even an alien civilisation based on Alpha Centauri would calculate exactly the same numeric value for these constants. Hence, the numeric values of these constants are particularly interesting as they seem to capture some deep truth about the workings of the universe.

It was Albert Einstein who first noticed a particularly interesting feature about the dimensionless constants: their values tend to be close to one. This tendency has proven to be a useful tool for physicists. As John Barrow explains in his book *The Constants of Nature*: "In every formula we use to describe the physical world, a numerical factor appears ... which is almost always fairly close in value to 1 and they can be neglected, or approximated by 1, if one is just interested in getting a fairly good estimate of the result."

Why should it be the case that the values of the dimensionless physical constants tend to be close to one? Well, if we had complete knowledge about the underlying physical mechanism, we might imagine that we could write down a mathematical formula that accurately describes that mechanism, and produces the value of the constant as a result. In that case, it actually becomes difficult to imagine a simple mechanism which could produce values which are many orders of magnitude greater than one. This is because most of the fundamental mathematical constants we would use in our formula (such as pi) also tend to have values very close to one. In that case, how could a simple mechanism produce a numeric value of the order of 10^{40}, for example? It is not easy to imagine how that could happen.[9]

There is another very good reason for expecting the values of the dimensionless constants to be close to one: it is because this would represent a situation which is considered *natural*. As described in Chapter One, we might imagine the universe as being described by Martin Rees's six dimensionless numbers, and we might also imagine those

---

[9] One possibility might be via an *exponential* mechanism. As an example, place a grain of rice on the first square of a chessboard. On each successive square, double the number of rice grains. You will end up with $2^{64}-1$ (note: all small numbers in that expression) grains of rice on the chessboard, which is equal to the immensely large number 1.8×10^{19}.

numbers as being the settings on six dials. It would then appear that there is a considerable amount of "arbitrariness" in the universe, in that those dials might well have been set to other values than the observed values. This seems to indicate that the universe could take other forms based on other settings of the dials. In that case, no particular form of the universe appears to be favoured over any other form, and we are left with the question as to why the universe takes the form it does.

However, there is one setting of the dials which seems to be special. If a dial is set to the value 1.0, then we can effectively eliminate that dial without affecting the behaviour of the system (because multiplying or dividing by 1.0 leaves the original number unchanged). The following diagram shows a system in which a dial set to the value 1.0 can be removed without affecting the output of the system:

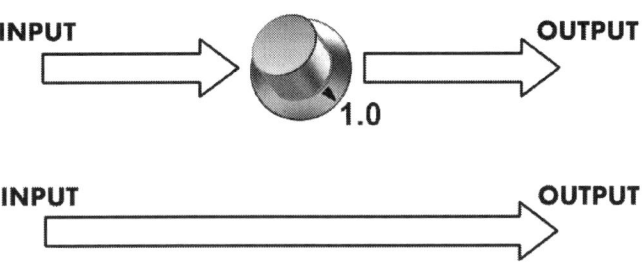

So a system which has all its arbitrary "dials" set to 1.0 could potentially have all its dials removed. The resulting system would not only be simpler, it would also have no arbitrariness – the system could not be modified (as it has no dials). Such a system would then be called *natural*. It represents the system in its unmodified state, with no tinkering. This also represents the situation in which it is not possible to modify the system, that is, if the state of the system arises from logical necessity. As Einstein said: "What I am really interested in is whether God could have made the world in a different way; that is, whether the necessity of logical simplicity leaves any freedom at all."

This principle that we should expect the values of the dimensionless constants to be close to one is called *naturalness*. It is a principle which is rather controversial at the moment as some physicists have raised doubts about the range of applicability of the principle. However, I think it is hard to argue with the very general description I have presented here. If we pursue naturalness, then we are eliminating arbitrariness from our solutions. In a completely natural universe, the form of the universe would arise from the structure of our equations, not the setting of arbitrary constants.

It might be the case that the state of the universe does not arise as a logical necessity, in which case we might be forever left with the question of why the universe takes the form it does. In that case, the only answer to why the constants take the values they have would be "that's just the way it is". That rather unsatisfactory answer will always encourage the proponents of the strong anthropic principle to suggest that parallel universes are the solutions to all our problems. But until we reach that end-of-the-road scenario, naturalness seems to be a valuable guide.

I don't think we should be giving-up on naturalness yet.

The naturally-flat universe

As an example of how a natural solution avoids any arbitrariness, let us consider the apparent fine-tuning of one of Martin Rees's "six numbers". In Chapter One it was explained how the value of Ω, the Greek letter omega, represents the ratio between the total amount of mass and energy in the universe, and the critical density of the universe. The value of Ω appears very close to one, which results in a flat universe. This was described as being another example of fine-tuning: if the value of omega was much less than one, the universe would have expanded too fast to allow galaxies and stars to condense, so life could not have evolved. Conversely, if the value of omega was much greater than one, the universe would have collapsed into a "Big Crunch" before stars could have formed.

However, on the basis of our discussion of naturalness, we can see that a value of 1.0 for omega would be a natural value, a value that requires no fine-tuning. It would be the value we should expect. The mystery then becomes the mystery of why our theories do not predict that value. Perhaps there is no fine-tuning – the failing might be in our theories to accurately capture the behaviour of Nature.

Without going into detail, my second book presented a hypothesis of modified gravity which resulted in a naturally-flat universe, and hence an apparent value of 1.0 for omega. There was no arbitrariness in the theory, no adjustable parameters – and no need for inflation. It would be an example of the power of the principle of naturalness.

Let us now examine some more of the deepest fine-tuning problems. We will need to consider our most accurate current theory of fundamental physics …

5
QUANTUM FIELD THEORY – PART ONE: EVERYTHING IS FIELDS

We are often told that quantum mechanics is a "revolutionary theory" and is the "the most successful theory in the history of physics", representing "our best theory of very small scales". However, as we shall see in this chapter, quantum mechanics cannot be a complete theory of reality at the fundamental level. We will find that a new theory is required.

We will see that this new theory has implications not only because it represents a more complete picture than quantum mechanics, but also because it completely reshapes our view of reality – in a way which is not widely realised or appreciated. We will see that this new theory is far more "revolutionary" than quantum mechanics ever was. We will see that this new theory forces us to give up our notion of a universe made of particles and instead consider a universe composed entirely of *fields*, which we can never directly feel or observe. These fields are continuous, and stretch from one end of the universe to the other. They are about as far from our intuitive notion of reality as it is possible to imagine.

However, we will also see that when we accept a universe made of fields and not particles, then concepts which seem so confusing in quantum mechanics suddenly start to make a lot more sense.

So, welcome to the theory called *quantum field theory*.

All of this might come as something of a shock, and I suspect it is not generally realised by the general popular science audience. However, while quantum field theory (generally known as QFT) might sound shocking and revolutionary, the accuracy of QFT has been thoroughly confirmed experimentally. QFT is now regarded as being our premier theory of fundamental physics.

As Roger Penrose says in his book *The Road to Reality*: "Quantum field theory constitutes the essential background underlying the Standard Model of particle physics, as well as practically all other physical theories that attempt to probe the foundations of physical reality. In fact, QFT appears to underlie virtually all of the physical theories that attempt, in a serious way, to provide a picture of the workings of the universe at its deepest levels."

QFT may be mind-altering, and revolutionary, but it is also the accurate description of how the universe works.

QUANTUM FIELD THEORY – PART ONE

A brief refresher on quantum mechanics

The material presented in this chapter will be heavily-based on material contained in my fifth book, which was an introduction to particle physics and quantum mechanics. It is highly-recommended that you read my fifth book in order to get sufficient background for the material presented in this chapter. As I do not like to repeat material covered in my earlier books, I am afraid the phrase "As explained in my fifth book …" is going to appear rather often in this chapter.

However, a very brief refresher on some of the concepts of quantum mechanics will be presented here – even though they have already been covered in greater detail in my fifth book.

According to quantum mechanics, before we observe some property of a particle, for example, its position or spin, we must consider the particle to be in a strange *superposition* state of all possible particle properties. The classic example of the double-slit experiment is often presented in which a particle appears to travel through two slits **at the same time** before making a mark on screen.

So that is the situation **before** we observe the particle. But **after** we observe the particle, we find the value of the observed particle's property must take only one of a number of certain allowed values. These allowed values are called *eigenvalues*, and the state of the particle must then lie in one of the allowed *eigenstates*.

This whole quantum measurement process can be described mathematically, or by the equivalent graphical description – which we shall now consider. The following diagram shows an extension to the double-slit experiment in which another slit is added (to convert it into a triple-slit experiment). In the diagram, the quantum state of the particle is shown by a vector (the dotted arrow).

The diagram shows the situation before measurement, i.e., before the particle hits the screen. Before measurement, you will see that the quantum state of the particle is in a superposition state: a combination of all three possible slits. Hence, it appears as though the particle is travelling through all three slits:

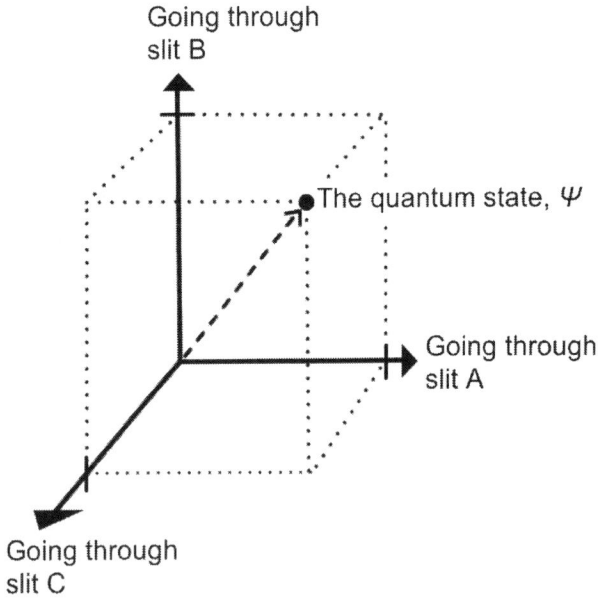

This form of graphical representation in which the quantum state is depicted as a vector is called a *Hilbert space*.

So what happens after we make a measurement? In other words, what happens when the particle hits the screen and reveals its true position? In that case, the particle leaves its superposition state and takes a single well-defined state. As an example, when we try to detect the position of the particle in our triple-slit experiment, we find it only passing through one slit. As you will see in the following diagram, this is represented by a rotation of the quantum state vector from its previous superposition state to one of the well-defined single states (in this case, going through slit B):

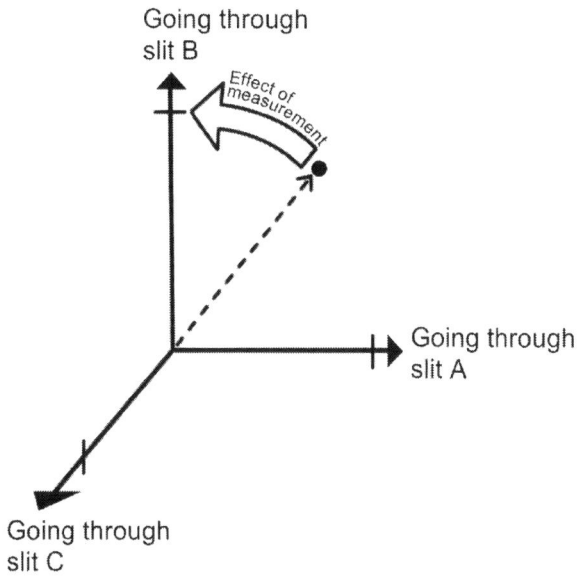

These three slits through which the particle can pass would then represent the eigenstates of the system – the only allowable states we will find when the observe the position of the particle.

Mathematically, this rotation of the quantum state is achieved by the application of an *operator* (just denoting the application of a mathematical operation).

Which brings us to the end of this very brief refresher on quantum mechanics.

The Dirac revolution

The highly-introverted physicist Paul Dirac is regarded as one of the key physicists of the 20th century. If anything, Dirac's reputation has only increased in the years since his death, and his work seems more central than ever.

Legend has it that Dirac's great moment of inspiration came in 1928 when he was staring into the fireplace at Cambridge University. Dirac realised how it was possible to combine quantum mechanics with special relativity. The result was that Dirac derived his famous equation for an electron which is known as the *Dirac equation* (the derivation of which was presented in my fifth book, together with more insights about the life and work of Paul Dirac).

The Dirac equation made some extraordinary predictions. It predicted the spin of an electron, which was quickly realised to be a perfect match with experiment. But, most extraordinary of all, the Dirac equation predicted new types of particles which we now know as *antimatter*. It is now known that all particles have an antimatter equivalent.

It was the discovery of antimatter which was to have such a revolutionary effect in quantum mechanics. If a particle of matter meets its antimatter equivalent then both particles disappear in a flash of energy (potentially quite destructive energy). The reverse process is also true: if enough energy can be confined to a small space then a particle can be created together with its antimatter equivalent (a process known as *pair production*). It is possible to make particles! This means that the number of particles in the universe is not a fixed number.

This development is described by Roger Penrose in his book *The Road to Reality*:

> The key property of an antiparticle is that the particle and antiparticle can come together and annihilate one another, their combined mass being converted into energy, in accordance with Einstein's $E=mc^2$. Conversely, if sufficient energy is introduced into a system then there arises the strong possibility that this energy might serve to create some particle together with its antiparticle. Thus, our relativistic theory certainly cannot just be a theory of single particles, nor of any fixed number of particles whatever. Indeed, according to a common viewpoint, the primary entities in such a theory are taken to be the quantum fields, the particles themselves arising merely as "field excitations".

Pay particular attention to Roger Penrose's last sentence which suggests that the "primary entities" are the fields – not the particles. We will be returning to this idea shortly.

Fock space

At this point, our theory of quantum mechanics hits something of a problem. Quantum mechanics is, essentially, a more complete update of classical mechanics (the mechanics of Newton). Hence the use of the word "mechanics". So, like classical mechanics, quantum mechanics is basically a theory of how things move, the "things" in this case being particles. To see the problem, let's consider an example from classical mechanics.

Imagine you are pushing a heavy wheelbarrow up a hill. If we want to analyse the motion, and calculate how hard we have to push, we can use classical mechanics and Newton's laws of motion to calculate the forces involved. We can therefore create a perfectly accurate mechanical model of the situation.

But, all of a sudden, something very surprising happens: the wheelbarrow completely disappears into thin air. This poses a serious problem for our mathematical model: there is nothing in Newton's laws of motion which says anything about objects disappearing. Likewise with quantum mechanics, which is just an update on classical mechanics: there is no way in quantum mechanics to model objects which just suddenly decide to disappear.

As Leon van Dommelen says in his excellent web-based book *Quantum Mechanics for Engineers*: "The quantum formalism in this book cannot deal with particles that appear out of nothing or disappear. A modified formulation called quantum field theory is needed."[10]

So when Paul Dirac revealed that it was possible for particles to appear and disappear, this went beyond the capabilities of the Hilbert space representation. A new mathematical representation is needed for quantum field theory. We have to move from the Hilbert space representation to the *Fock space* representation.

Fock space looks rather similar to Hilbert space in that the coordinate axes represent eigenstates of the system (the allowable states which can be observed after measurement). Let us consider the Fock space for a number of identical particles contained in a box. The eigenstates on each axis of

---

[10] *Quantum Mechanics for Engineers* by Leon van Dommelen, http://tinyurl.com/qmforengineers

the corresponding Fock space will represent one possible combination of the particles in the box.

As the axes of the Fock Space represent **all** possible combinations of the particles in the box, it is necessary to consider the situation when some particles appear and disappear. Hence, the axes of Fock space can even represent different numbers of particles, as shown in the following diagram of a particular Fock space:

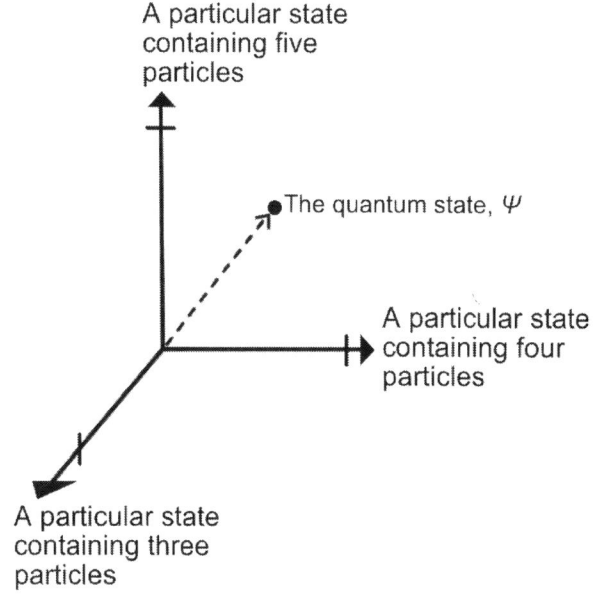

Note that each axis of the Fock space can represent the case of a different numbers of particles in the box (five particles, four particles, etc.).

A bit of terminology for you: when we move from classical mechanics to the Hilbert space representation it is called *first quantisation*. Then when we move from the Hilbert space representation to the Fock space representation (the QFT representation) it is called *second quantisation*. A crucial point is that in the Hilbert space representation, not every situation can be represented (e.g., when the wheelbarrow disappears), whereas Fock space can mathematically represent **absolutely everything that can possibly happen.** This is a crucial feature of second quantisation in QFT, and we will be returning to consider this feature in the next chapter.

Looking at the axes of the Hilbert space and the Fock space, it is clear that there is a similarity between the two representations, with different eigenstates on each axis. And, just as in the Hilbert space representation, it is possible in the Fock space representation to apply different operators to rotate the current quantum state to different eigenstates (which would represent the state of the system when we open the lid of the box and count the particles).

As an example, the following diagram shows the result of a *creation operator* when applied to a box containing four particles:

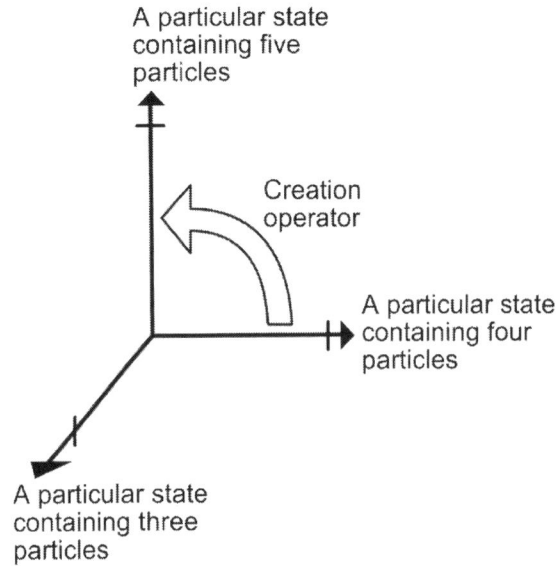

You will see that the result of applying the creation operator has been to rotate the quantum state of the system from a state which has four particles to a state which has five particles. Hence, this represents the creation of a particle.

Similarly, it is possible to define an *annihilation operator*:

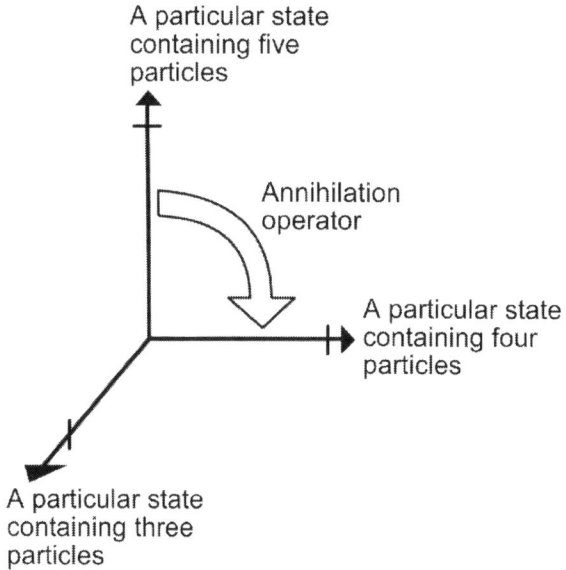

You will see that the result of applying the annihilation operator has been to rotate the quantum state of the system from a state which has five particles to a state which has four particles. Hence, this represents the disappearance of a particle.

You could apply the creation operator to the quantum state repeatedly, each time adding a new particle inside the box (and generating a new quantum state in the process). Or you could apply the annihilation operator repeatedly, each time removing a particle from the box (until there were no more particles left to remove).

Everything is fields

We have to be careful when applying these creation and annihilation operators, because we have to make sure we do not do something which is forbidden by Nature. This is because all particles can be categorised into one of two groups: they are either *bosons* or *fermions*. Bosons like being in the same state (as described in my fifth book, this is due to the wavefunction of a boson being *symmetric*). As an example, a beam of light is composed of many billions of photons in the same state (photons are bosons). It is therefore possible to use the creation operator to create as many bosons in the same state as you wish. It is then said that those particles are *indistinguishable*.

However, two fermions cannot exist in the same state due to the *Pauli exclusion principle* (as described in my fifth book, this is due to the wavefunction of a fermion being *antisymmetric*). So we cannot use the creation operator to create two fermions (for example, two electrons) in the same state. For this reason, it is said that those particles are *distinguishable*.

However, in the Fock space representation, all we see is numbers of particles on each axis. In other words, in Fock space there is nothing to distinguish one particle (fermion or boson) from any other particle. This is a crucial point: in Fock space, and therefore in QFT, **all** particles (of the same type) are indistinguishable.

What is implied by the fact that all particles are identical in QFT? It has been suggested that it implies something extraordinary: there are not multiple particles, there is only the one field. As an example, **all electrons are identical because they are not actually separate objects, they are**

merely different parts of a single electron field which covers the entire universe.

This principle is described well by Leon van Dommelen:

> *There is a field of electrons, there is a field of protons (or quarks, actually), there is a field of photons, etc.*
>
> *Some physicists feel that is a strong point in favor of believing that quantum field theory is the way Nature really works. In the classical formulation of quantum mechanics, the (anti)symmetrization requirements under particle exchange are an additional ingredient, added to explain the data. In quantum field theory, it comes naturally: particles that are distinguishable simply cannot be described by the formalism.*

This principle is also described on the Wikipedia page for quantum field theory, under the section "Physical meaning of particle indistinguishability". Again, note that the crucial factor in QFT is that all particles (of the same type) are indistinguishable:[11]

> *The second quantization procedure relies crucially on the particles being identical. From the point of view of quantum field theory, particles are identical if and only if they are excitations of the same underlying quantum field. Thus, the question 'why are all electrons identical?' arises from mistakenly regarding individual electrons as fundamental objects, when in fact it is only the electron field that is fundamental.*

---

[11] http://en.wikipedia.org/wiki/Quantum_field_theory

So it is now believed that we live in a universe in which everything that exists is a field. Every particle is just a part of a single field for that particle type. Every field stretches the length and breadth of the universe.

How does the single field sometimes appear like particles? It is because particles emerge in the field like waves in the ocean – they are not actually separate from the field, just like a wave is not separate from the ocean. These particles emerge when enough energy is transferred to a small region of the field, like a rock dropped into the ocean.[12] Particles are often referred to as "excitations" of the field (see Roger Penrose's earlier quote). As an example, this is how the Higgs boson was generated from the Higgs field by the high-energy collider at the LHC.

Sean Carroll has an excellent talk on YouTube filmed at Fermilab called *Particles, Fields, and the Future of Physics* in which he emphasises this idea that everything is a field: "That particle stuff is overrated. Actually, everything is made of fields. You don't need to separately talk about matter made of particles and forces made of fields. All you need are fields."

Here is a link to Sean Carroll's talk:

http://tinyurl.com/seancarrolltalk

Here is another quote from Sean Carroll's talk. This is certainly a guiding principle behind the writing of this book:

---

[12] In order to create a particle of mass m, sufficient energy must be put into the field according to the formula $E=mc^2$.

QUANTUM FIELD THEORY – PART ONE

To working physicists, quantum field theory is the most important thing we know. But when we talk about physics to non-physicists, when we popularise it, we never mention quantum field theory. We talk about particle physics, we talk about relativity, we talk about quantum mechanics. Heck we even talk about string theory and the multiverse and the anthropic principle. But we think quantum field theory is too much to bother about.

Another populariser of the "Everything is Fields" philosophy is the retired physicist Rodney Brooks. Brooks is no longer working in academia, but in the 1950s he was taught by Julian Schwinger who was one of the founders of QFT, and Brooks has made it his retirement project to raise awareness and understanding of QFT in the general public. He is very much an evangelist of the idea that fields are everything, and everything is fields: "No particles, only fields".

Brooks has created a video in which he explains his views on QFT:

http://tinyurl.com/rodneybrooks

Brooks also appears to be on a mission to increase awareness of his tutor, the Nobel prize-winning Julian Schwinger. Schwinger's field-based approach to QFT lost in the popularity stakes to the particle-based approach to QFT which was easier to understand and calculate and was popularised by the more charismatic Richard Feynman (who we will be considering in the next chapter). Here is a photograph of Schwinger, who undoubtedly deserves more recognition as one of the most important figures of 20th century physics:

In his video, Brooks mentions how Richard Feynman said it was impossible to understand the counter-intuitive behaviour of quantum mechanics. Brooks then explains how an understanding of QFT can explain effects which seem so puzzling in quantum mechanics:

> *In quantum field theory, the uncertainty principle becomes trivial. A field is spread-out. An electron is a particle which might be here, it might be there, might have this momentum, might have that momentum – that doesn't make sense. Feynman is correct. But a field that is spread out is not just 'could be here' or 'could be there' – it's here, and here, and here. So that problem, for example, is resolved.*

I would agree with Brooks's explanation of the uncertainty principle: uncertainty about position makes a lot more sense when you consider fields rather than point-particles. Also, quantum entanglement – a connection between particles over great distances – also makes more sense when we consider fields extended through space rather than point-particles. And at the 28-minute mark of his aforementioned video, Sean Carroll also explains how a field (or wave) approach can provide a simple explanation for the interactions between particles.

However, so far we have only considered isolated particles in fields in which particles can appear and disappear (so-called *free fields*) but do not interact with each other. In the next chapter we will consider the significant problems which arise when we introduce the possibility of interactions between particles.

6

QUANTUM FIELD THEORY – PART TWO: THE FEYNMAN APPROACH

We have seen in the previous chapter how quantum field theory appears to describe the lowest level of reality. But it was not this field-based picture which made QFT such a success. Instead, it is a particle-based version of QFT developed by Richard Feynman which dominates QFT calculations. We will be seeing that this approach involves generating a large number of *Feynman diagrams* which describe particle interactions. This version of QFT has been proven to be incredibly accurate, and a powerful tool for physicists. But we will also see that the mathematics behind this approach has been criticised for lacking rigour and elegance.

Julian Schwinger certainly was rather unimpressed by this contrasting approach to QFT. Schwinger felt Feynman diagrams encouraged students to think in terms of particles instead of fields. However, Schwinger also had to reluctantly admit: "Like the silicon chips of more recent years, the Feynman diagram was bringing computation to the masses."

The totalitarian principle

In the previous chapter, it was explained how any successful theory of QFT has to be able to capture the possibility that "absolutely anything can happen", including particles being able to appear and disappear. This principle lies at the heart of QFT, and this behaviour was succinctly described by Murray Gell-Mann in his so-called *totalitarian principle*: "Everything not forbidden is compulsory".

The totalitarian principle was certainly the guiding principle in the approach to QFT which was developed by Richard Feynman. As mentioned earlier, Feynman's approach was particle-based rather than field-based. To understand Feynman's approach, consider a particle moving in space from a point A to a point B. In classical mechanics, this is obviously straightforward: we can see a large object – a train, for example – as it proceeds on its journey between two points. For particles, however, things are not so clear-cut. For a particle moving from point A to point B, we can only observe the particle when it sets off on its journey (at point A) and when it arrives at its destination (at point B). We have no information about what the particle is doing in-between those two points. Crucially, we can have no information about the **route** the particle is taking between those two points. We are **fundamentally forbidden** from knowing the route. This is basically the principle behind the double-slit experiment: we do not know which slit the particle passes through.

But – you might argue – why can't we just observe the particle in between the two points, thus revealing its path? Well, the only way we could do that would be to arrange for the particle to collide with a secondary particle, for example, to be observed by a photon of light. But if we create an interaction in that manner, then we are inevitably altering the path of the particle under observation: we have modified the experiment we wished to observe. There is no way round it: we are fundamentally forbidden from being certain of the path the particle is taking in order to get from point A to point B.

So Richard Feynman considered the problem of a particle travelling between two points and he applied the totalitarian principle quite literally. Feynman stated that in order to calculate the probability of a particle travelling between two points, we have to sum the probabilities of the particle travelling along **all possible paths** between those two points. And when Feynman said "all possible paths" he most certainly meant **ALL** possible paths. We even have to consider the possibility of the particle taking an apparently crazy route – such as detouring via the Moon, or even the Andromeda galaxy!

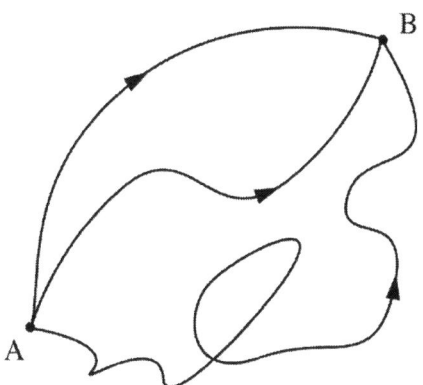

This approach is called the *Feynman path integral* (an "integral" is just a summation of lots of smaller elements, like the summing of the small path elements along the curve). To be precise, if we consider the particle as moving in some field of force, then we have to integrate the

Lagrangian along the path (the Lagrangian was described in my second book, and is equal to the particle's kinetic energy minus its potential energy due to the field of force). The integration process involves dividing the path up into infinitely small pieces and then adding all of those pieces together.

However, a problem arises when we have to add all of the possible paths together as there are infinitely many possible paths. We now see the problem of infinities in QFT: how can we possibly add together an infinite number of objects and still arrive at a finite result? This problem arises because — as was stated earlier — in QFT we have to consider **absolutely anything that can possibly happen.** As we shall soon see, the challenge for any computational technique in QFT is to avoid the infinities which are introduced.

Fortunately, for the case of the Feynman path integral, there is a relatively straightforward reason why the final result is not infinite. Remember, we are dealing with quantum particles, and the position of a quantum particle is always governed by probability. In particular, when the particle travels between two points its position is determined by a wave — a *wavefunction* — and the probability that the particle is in a certain position can be calculated from the square of the wavefunction at that position.

So a quantum particle acts like a wave. And, like any wave, it has an associated angular *phase*. If we add two waves and the phases are the same, then the result will be a large value (this is called *constructive interference*). However, if the phases of the two waves are opposite (pointing in opposite directions) then the waves will cancel (this is called *destructive interference*).

Fortunately, in the Feynman path integral, the more crazy paths — such as travelling to Andromeda or travelling to the Moon — tend to point in various unrelated random directions and so tend to cancel each other due to destructive interference. We are left with the most direct route from point A to point B being the route taken by the particle. This would be the same classical route we would expect a large object to take, such as a ball. As Brian Greene says in his book *The Elegant Universe*:

> *No matter how absurd nature is when examined on microscopic levels, things must conspire so that we recover the familiar prosaic happenings of the world experienced on everyday scales. To this end, Feynman showed that if you examine the motion of large objects – like baseballs, airplanes, or planets, all large in comparison with subatomic particles – his rule for assigning numbers to each path ensures that **all paths but one cancel each other out** when their contributions are combined. In effect, only one of the infinity of paths matters as far as the motion of the object is concerned. And this trajectory is precisely the one emerging from Newton's laws of motion.*

In his book of lecture notes called *QED: The Strange Theory of Light and Matter*, Richard Feynman considered a photon reflecting off a mirror. We would intuitively imagine the path taken by the photon to be the same as that of a bouncing ball, with the angle of incidence being equal to the angle of reflection off the mirror. However, Feynman described how the actual quantum process underlying the behaviour of the photon is far from intuitive.

The following diagram shows a photon of light being reflected off a mirror. The photon is emitted at point X, is reflected off the mirror (shown by the long thin rectangle along the bottom of the diagram), and is detected at point Y:

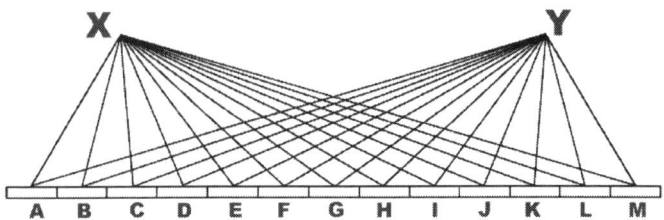

You will see that Feynman divided the long rectangular mirror into small segments and denoted those segments by letters, from A to M. The mid-point of the mirror is segment G. Intuitively, we would imagine the photon simply being emitted at point X, being reflected off segment G, and being detected at point Y. We would certainly not expect any of the other segments of the mirror – from A to F, or from H to M – to have any involvement in the reflection process. As Feynman says in his book:

> *We would expect that all the light that reaches the detector reflects off the middle of the mirror, because that's the place where the angle of incidence equals the angle of reflection. And it seems fairly obvious that the parts of the mirror out near the two ends have as much to do with the reflection as with the price of cheese, right?*

However, Feynman then proceeds to apply his path integral technique to this example, and surprisingly reveals that even segments of the mirror which are far away from the midpoint play an important role in defining the reflection. Remember, the Feynman path integral says that we have to consider the photon taking **all possible routes** from point X to point Y. This means that we have to consider the photon also reflecting off segments which are far away from the middle of the mirror, such as segment A and segment M. In fact, we have to consider the photon reflecting off **all** the segments.

Fortunately, though, when we add up all the probabilities of all the different possible paths, we once again find that the apparently crazy paths – reflecting off the segments near the ends of the mirror – tend to cancel out. We are left with the route of the photon reflecting off the middle segment G.

However, I hope this has shown that effects which we intuitively consider to be common sense – such as a bouncing ball, or a reflecting ray of light – are actually products of quantum effects which involve taking **all possible routes**.

Feynman diagrams

Richard Feynman's greatest achievement was the joint development of the theory of *quantum electrodynamics* (also known as QED) which is the quantum field theory of the electromagnetic force. As mentioned in the previous section, Feynman wrote a very accessible book based on his lecture notes on quantum electrodynamics called *QED: The Strange Theory of Light and Matter*. As Feynman says in his book in his usual conversational style:

> *What I'd like to talk about is a part of physics that is **known**, rather than a part that is unknown. People are always asking for the latest developments in the unification of this theory with this theory, and they don't give us a chance to tell them anything about one of the theories that we know pretty well. They always want to know things that we don't know. So, rather than confound you with a lot of half-cooked, partially analyzed theories, I would like to tell you about a subject that has been very thoroughly analyzed. I love*

this area of physics and I think it's wonderful: it is called quantum electrodynamics, or QED for short.

My main purpose in these lectures is to describe as accurately as I can the strange theory of light and matter – or more specifically, the interaction of light and electrons.

As Feynman said in the previous quote, his particle-based approach to quantum field theory meant that his version of QED was a theory of the interactions between electrons and photons (which are the field particle of the electromagnetic force). As a result, in the examples in this section we will be considering the interactions between electrons and photons.

Perhaps Feynman's most significant contribution to quantum field theory came from his introduction of *Feynman diagrams* which were based on his path integral formulation. Feynman diagrams are simple representations of the interactions between particles. Once again, their ability to represent **absolutely anything that can possibly happen** makes them perfect for calculations in quantum field theory.

A Feynman diagram is a spacetime diagram, with space along one axis, and time along the other axis. Even though space is three-dimensional, only one dimension of space is included on the diagram.

The following diagram is an example of a Feynman diagram. Note that time increases along the horizontal axis (they are sometimes drawn with time on the vertical axis). The diagram shows an electron coming in from the left of the diagram, and colliding with a photon which changes the path of the electron:

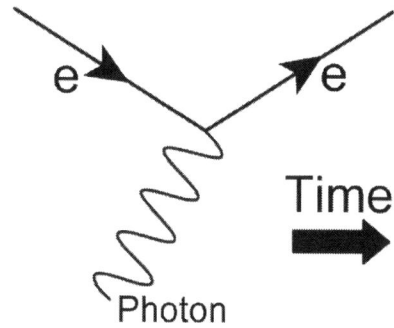

You will notice that the electron is denoted by a straight line (all fermions are denoted by straight lines) while the photon is denoted by a wavy line (all bosons are denoted by wavy lines). Each line – together with its associated mathematical probability – is called a *propagator*.

This very simple Feynman diagram is called the *minimal interaction vertex* (a *vertex* is the point where the lines meet on the diagram). Rather wonderfully, we can create all of the Feynman diagrams for the electromagnetic force just by glueing together these minimal interaction vertex diagrams – as we shall soon see. As David Griffiths says in his textbook *Introduction to Elementary Particles*: "All electromagnetic phenomena are ultimately reducible to this elementary process." Bruce Schumm makes the same point in his book about particle physics *Deep Down Things*: "It is the general electron-photon vertex that we should look to as the fundamental component of the electromagnetic force."

So remember what the minimal interaction vertex looks like: it is the junction of two electron paths and one photon path. You will be recognising this arrangement again soon.

To show the amazing power of the minimal interaction vertex, let us consider what happens when we rotate the diagram. Relativity tells us that we can treat time as just another dimension – just like the three dimensions of space.

So there is complete freedom to rotate Feynman diagrams (converting space into time, and vice versa) and the resulting interaction will still be valid. As an example, the following diagram shows the previous minimal interaction vertex rotated 90° anticlockwise:

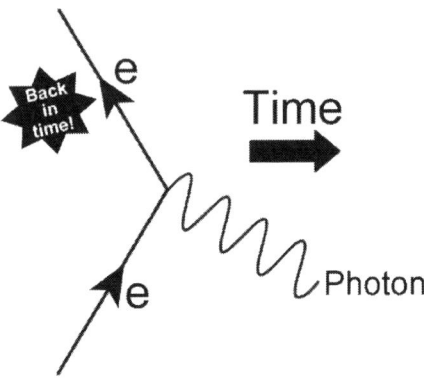

There is now a remarkable interpretation possible for this diagram. You will see that it appears that the electron is travelling backward in time on the second half of its path (denoted by the star on the diagram). Also, if we temporarily ignore the arrows on the paths of the electrons, we could interpret the left-hand side of the diagram as representing two electrons coming together and annihilating each other – producing a photon as the result. We could therefore interpret this as an electron and a *positron* (the antimatter equivalent of the electron) annihilating each other and producing "pure energy". In which case, that would mean a positron is actually an electron travelling backward in time (as shown in the star on the diagram). It was this definition of a positron which Feynman often used in his diagrams.

QUANTUM FIELD THEORY – PART TWO

Let us now examine how Feynman diagrams can be used in quantum field theory calculations.

Firstly, in the following diagram, let us simply consider two separate electrons moving through spacetime and ending up at two different points. The simplest way that this can happen is shown here:

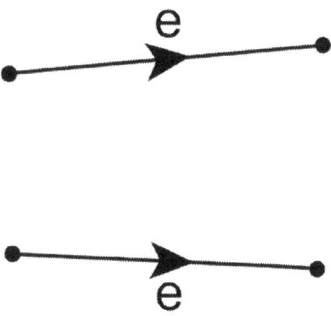

What we are interested in is calculating the probability that this situation shown actually happens (i.e., both electrons move as shown in the diagram). Fortunately, we know precisely how to do this as we can use the Feynman path integral technique which was explained earlier in this chapter. If you remember, the path integral technique gives the probability of a particle moving between two points, so it is just what we need here.

Then, to calculate the complete probability of the situation shown in the previous diagram, we need to calculate the probability of the first electron moving along its path, and then **multiply** that probability by the probability of the second electron moving along its separate path (according to basic probability theory, we have to multiply the probabilities together because **BOTH** event one **AND** event two have to happen to provide the correct result).

But there is another way that the two electrons could get to their endpoints. As shown in the following Feynman diagram, during their journey they could transfer a photon between each other, modifying their trajectories in the process:

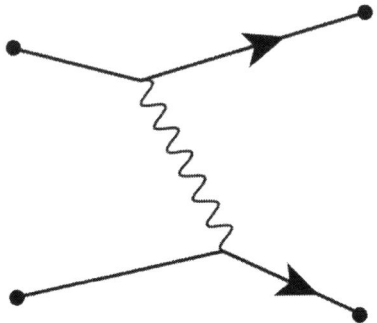

How do we know this is a valid interaction? We know it is valid because we can create this Feynman diagram by glueing together two minimal interaction vertices ("vertices" being the plural of "vertex"):

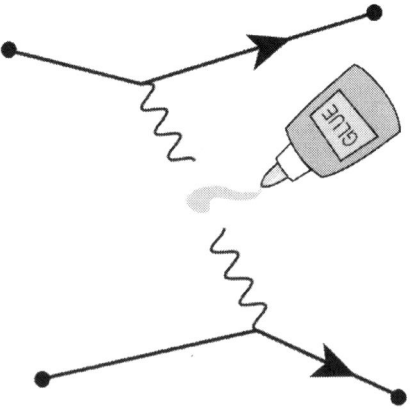

If you remember, a minimal interaction vertex is a junction of two electron paths and one photon path. So you can see that the previous diagram represents two minimal interaction vertices glued together. Everything fits together like Lego! And the golden rule is that if you can create a Feynman diagram by glueing together minimal interaction vertices, then it is a valid Feynman diagram, and you have to consider it in your calculations.

So that gives us two different ways for the electrons to move to their two separate endpoints. If we want to calculate the probability of the two electrons moving to their endpoints we have to **sum** the probabilities of these two separate diagrams (according to basic probability theory, we have to sum the probabilities because **EITHER** diagram one **OR** diagram two would provide the correct result).

But that's not the end of it. It is possible for the photon to convert into an electron/positron pair for a very short period of time. The electron and the positron can travel a short distance before annihilating each other to produce the photon again. This is shown by the loop in the photon's path in the following diagram:

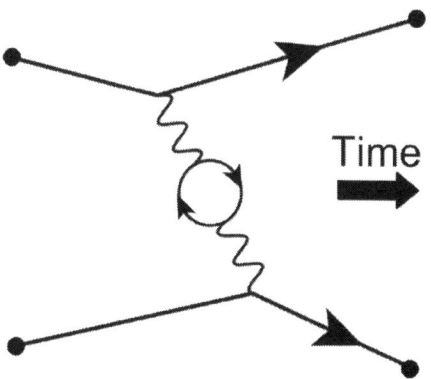

In the previous diagram, we can consider the loop as being made of the paths of two particles which are shown by the two arrows in the loop. You can see that one particle is travelling forward in time and is therefore an electron, whereas the other particle is travelling backward in time and is therefore a positron.

How do we know this new Feynman diagram is valid? Well, we know it is valid because we have merely added two new minimal interaction vertices, as shown in the following diagram:

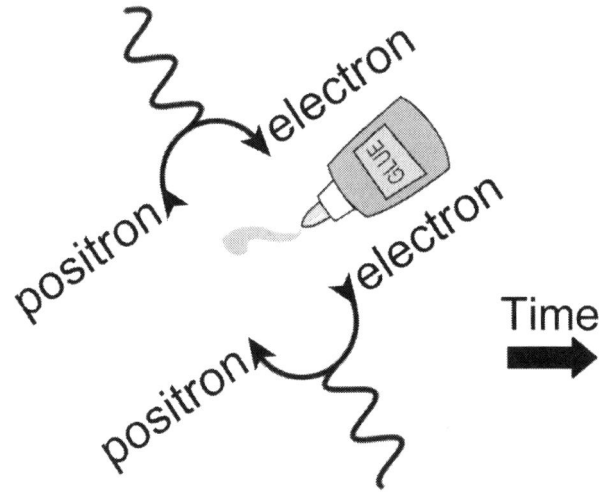

You can see that all we have done is glue together two more minimal interaction vertices (don't be confused by the curved lines rather than straight lines). Once again, everything fits together like Lego.

The electrons and photon in the previous diagram only appear in internal lines in the Feynman diagram. These particles which only appear on internal lines (particles which appear only to rapidly disappear again) are called *virtual particles*.

So now we have a total of three different Feynman diagrams to add together to find the probability of two electrons simply moving to two separate endpoints. Here are the three diagrams we must add:

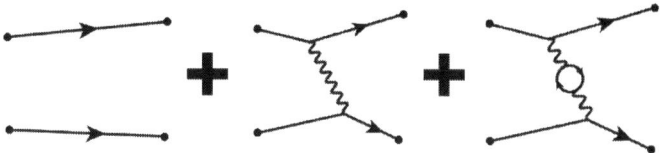

And it doesn't stop there! We can keep going forever, dividing lines, and adding new minimal interaction vertices. With each new diagram, our estimate of the probability of the interaction becomes more accurate. This repeated approach is called *perturbation theory*.

Unfortunately, it appears inevitable that we are going to end up with an infinity of Feynman diagrams which we have to add together just to calculate the probability of two electrons moving. Once again we end up with the problem of how to handle the infinities when making calculations in quantum field theory. How on earth can we add up an infinite number of Feynman diagrams and still get a finite answer?

The fine structure constant

As described earlier in this section, in order to calculate the probability of an entire Feynman diagram, we have to multiply **all** the probabilities contained within that diagram. We can calculate the probability of a particle moving along a path using the Feynman path integral technique. However, that method does not account for all the probabilities in the diagram: we also have to consider the probability of an electron emitting a photon. And we will now see that it is the particular value of this probability which saves us from the dreaded infinities entering our calculations.

Rather simply, the probability of an electron emitting a photon is just a fixed number. The number reflects the strength of the electromagnetic force. There are four fundamental forces, and they each have a characteristic number which reflects the strength of each force. The number for each force is called the *coupling constant*. For the example of the electromagnetic force, which interests us in particular, we will now see how this value is calculated.

Let us consider the strength of attraction between two electric charges a distance r apart. According to *Coulomb's law*, the force, F, is given by:

$$F = \frac{e^2}{4\pi\varepsilon_0 r^2}$$

where e is the elementary charge carried by a single electron (1.6×10^{-19} coulombs), and ε_0 is the permittivity of free space (8.85×10^{-12} farads/metre).

If you read my previous book then you will know that the r^2 term arises purely from the fact that there are three dimensions of space (the effect of a force dilutes as it extends into space according to an *inverse square law*). So the r^2 term is not specific to the electromagnetic force – it applies to all forces. Hence, in our analysis we should remove the term: we are only interested in the strength of the force in terms of fundamental constants.

As a result, we find the strength of the electric force is then characterised by:

$$\frac{e^2}{4\pi\varepsilon_0}$$

If we consider the units (dimensions) of this expression then we find it has units of energy times distance. As explained in Chapter Four, we do not want to have a term with units as its numerical value will change if we choose different units. We want to find a dimensionless number (i.e., a plain number without units). We can change this term into a dimensionless number if we divide the expression by the product of two fundamental constants, $\hbar c$, a term which also has dimensions of energy times distance ($\hbar$ is the *reduced Planck constant* which is equal to the Planck constant divided by 2π, and c is the speed of light).

We then end up with a mathematical expression describing the strength of the electromagnetic force which is called the *fine structure constant*, and is often denoted by the Greek letter α (alpha):

$$\alpha = \frac{e^2}{4\pi\varepsilon_0 \hbar c}$$

The fine structure constant is a famous number in physics and, if you substitute the correct values (given earlier) into the formula, you will find it has a value approximately equal to 1/137. The fact that this vital constant has such an interesting value has made it a target for numerologists who have attempted to derive its peculiar "137" value from various combinations of other constants. Most physicists would regard these attempts as being unscientific (randomly throwing numbers together and hoping to come up with the correct solution shows a lack of understanding).

What makes the fine structure constant so interesting from the point of view of our Feynman diagrams is that it represents the probability of an electron emitting a photon (hence it determines the electric force between two electrons). To be more precise, the fine structure constant represents the probability of a **pair** of electrons emitting and absorbing a photon (because an electron/photon interaction always requires two vertices: one electron to emit the photon, and a second electron to absorb the photon). But what really interests us is the probability of one vertex: a single electron emitting a single photon, so that value is provided by the **square root** of the fine structure constant (because when the value is then squared by multiplying it by itself due to the pair of interactions it will result in the fine structure constant).

Richard Feynman makes this point in his book:

There is a most profound and beautiful question associated with the observed coupling constant, the amplitude for a real electron to emit or absorb a real photon. It is a simple number that has been experimentally determined to be close to 0.08542455. My physicist friends won't recognize this number because they like to remember it as the inverse of its square: about 137.03597. It has been a mystery ever since it was discovered more than fifty years ago, and all good theoretical physicists put this number up on their wall and worry about it.

What makes this coupling constant so vital from the point of view of our Feynman diagrams is that its value is less than 1.0. If you remember, when we calculate the probability of a Feynman diagram we have to multiply all of the probabilities of all the interactions contained within that diagram. If the probability of an electron/photon vertex is just 0.085 then the overall probability of a Feynman diagram will reduce rapidly for each vertex we add.

This reduction in probability is shown in the next diagram. For Feynman diagrams with two electron/photon vertices there is only a probability of 0.7% that the event described by the diagram will actually occur. For Feynman diagrams with four electron/photon vertices there is only a probability of 0.005% that the event will occur (a less than one in ten thousand chance).

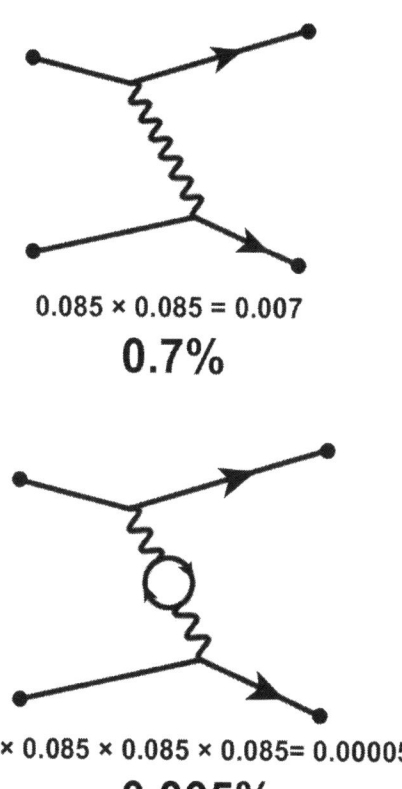

0.085 × 0.085 = 0.007
0.7%

0.085 × 0.085 × 0.085 × 0.085 = 0.00005
0.005%

As you can see from the previous diagram, because the value of the coupling constant is less than 1.0, this means that Feynman diagrams with multiple vertices will have a very low probability and can be safely ignored by our calculations. Hence, we do not have to add an infinite number of Feynman diagrams together in order to calculate a sufficiently accurate result: the value of the fine structure constant has saved us. Another infinity has been avoided!

However, this is not the case for all forces. The strong force has a coupling constant of 1.0, so this will not become progressively less important as more diagrams are added. Hence, perturbation theory using Feynman diagrams cannot be used to perform calculations involving the strong force.

Interestingly, this probability-based approach gives us another way of thinking about the "strength" of a force. A force with a larger coupling constant simply means there is a greater probability of more force-carrying bosons being produced – and that translates into a stronger force between particles. As Bruce Schumm says in his book about particle physics *Deep Down Things*: "The stronger the coupling, the more quanta are exchanged in a given interaction." Considering an analogy to skaters on ice throwing a ball between themselves, Schumm says: "A stronger force would correspond to a greater likelihood, in any given instant, that a ball would be exchanged between them."

QUANTUM FIELD THEORY – PART TWO

The Higgs mechanism

One of the fields which crosses the entire universe is called the *Higgs field*. The behaviour and structure of the Higgs field is notably different from the other fields. The Higgs field is responsible for giving mass to elementary particles.[13]

The way the Higgs field gives mass to fermions is rather different to how it gives mass to bosons. Firstly, let us consider how the Higgs field gives mass to fermions.

Unsurprisingly, there is a particle which is associated with the Higgs field (it was explained in the discussion of quantum field theory in the previous chapter how every field has an associated particle which can be produced by an excitation of the field). The particle associated with the Higgs field is the famous *Higgs boson*. As a fermion travels through the Higgs field (which permeates the entire universe) it interacts with a sea of Higgs bosons. The result is that the particle slows down as if it is being pushed through a vat of sticky molasses (another common analogy is of a popular person trying to cross a room in a crowded party). As mass is defined as "resistance to acceleration", we interpret this slowing as the particle gaining mass.

---

[13] For *composite* particles such as protons, i.e., particles which are composed of elementary particles grouped together, most of the mass arises from the binding energy which holds the elementary particles together – **not** the Higgs field (the binding energy is equivalent to mass via $E=mc^2$).

To understand how the Higgs field gives mass to bosons, let us first consider the difference between particles which have mass (*massive* particles) and particles which do not have mass (*massless* particles).

The theory of special relativity states that if we try to accelerate an object up to the speed of light, the mass of that object will progressively increase. At the speed of light, the mass of the object would become infinite. In other words, it is not possible for an object with mass to move at the speed of light. This is a principle which applies to all massive particles such as electrons: electrons have to move at less than the speed of light.

However, a photon is an example of a massless particle. And if a particle has no mass, then it has no mass to increase as the speed of the particle approaches the speed of light. It is therefore possible for a massless particle to move at the speed of light. After all, a photon is a particle of light, so it obviously has to move at the speed of light.

Put simply, all massless particles move at the speed of light, and all massive particles have to move at less than the speed of light.

This principle has implications for the spin of massive and massless particles. If we consider a particle in three-dimensional space, it is clear that there are three axes about which it can spin (the z-axis is coming out of the page):

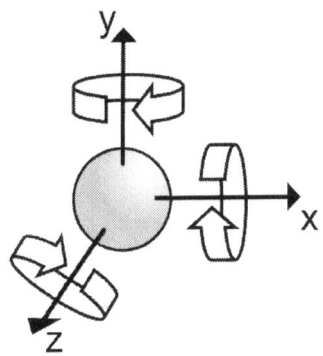

Each of these spin axes is called a *degree of freedom*, with each degree of freedom representing an independent way in which the particle can move. However, if we consider a massless particle, i.e., a particle which is moving at the speed of light, then we encounter a problem.

To understand the problem, consider the following diagram which shows a massless particle moving at the speed on light in the x direction. You will see from the diagram that if the particle is also rotating about either the y or z axis then there would always be a part of the particle which would be moving faster than the speed of light (the top semicircular arrow would be moving faster than light, while the bottom arrow would be moving slower than light):

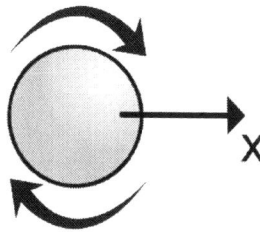

This is obviously forbidden, so the massless particle is forbidden from rotating about its y or z axis.

This is not a problem if the particle is rotating about its x axis because then the plane of rotation would be perpendicular to the particle's motion along the x axis. So a massless particle is allowed to rotate in a plane perpendicular to its direction of motion. On that basis, the following diagram shows a massless particle moving at the speed of light in the x-direction, while also rotating about that x axis:

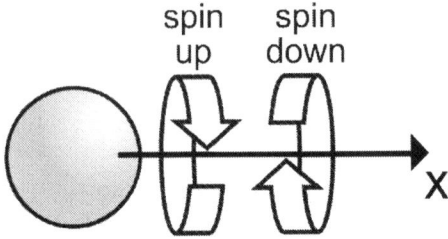

It is clear from the previous diagram that the particle can spin in either direction around the x-axis, and we call these directions *spin up* and *spin down* (as shown on the previous diagram). Hence, the massless particle has two rotational degrees of freedom. These two states are called the *polarization* of the particle, and it is the reason why light (which is composed of massless photons) has two directions of polarization. This is the basis for polarized sunglasses (one of the two directions of polarization – the horizontal glare from road surfaces – is filtered-out).

Now let us consider what happens when the massless particle interacts with the Higgs boson – the particle of the Higgs field.

When the massless particle combines with the Higgs boson, it gains one additional degree of freedom which is possessed by the Higgs boson. This additional degree of freedom becomes a freedom of motion along the x-axis, the direction in which the particle is moving. And if the particle has a freedom along the x-axis then it is not longer restricted to moving at only the speed of light. In other words, the particle might be said to be "slowing down". It has already been described how massive particles (particles with mass) move slower than the speed of light, so we interpret this "slowing" of the massless particle as the particle "gaining mass".

Hence, the Higgs field can give mass to otherwise massless bosons.

QUANTUM FIELD THEORY – PART TWO

The hierarchy problem

But the Higgs boson – which was detected for the first time by the LHC in 2012 – introduces an additional fine-tuning problem. The problem is associated with the measured mass of the Higgs boson.

The value of the Higgs mass cannot be calculated – it has to be measured experimentally. The LHC discovered the Higgs has a mass of approximately 125 GeV. Intuitively, though, particle physicists would have expected the Higgs mass to take a "natural" value, which turns out to be vastly heavier. As explained in Chapter Four, a natural mass would be one in which all the "arbitrary dials" were turned to 1.0. In which case, the mass would have to be calculated only from fundamental constants. The three truly fundamental constants are the speed of light, c, the gravitational constant, G, and the Planck constant, h. These three constants can be combined in only one way to produce a quantity which has the units of mass, and this is shown in the following formula:

$$m_P = \sqrt{\frac{\hbar c}{G}}$$

The value of this mass, m_p is called the *Planck mass*.

If you substitute the correct values into the formula for the Planck mass you will find you get a surprisingly heavy value: approximately 21 micrograms, the weight of an eyelash. So the measured Higgs mass value is vastly lighter in comparison, which is a surprise and a puzzle.

Another reason why the Higgs mass appears to be surprisingly light arises when we consider the effects of virtual particles temporarily popping into existence (virtual particles were considered earlier in this chapter). As described in the previous section, the Higgs is responsible for giving mass to particles, so, conversely, the Higgs can interact with any particle that has mass. As an example, the following diagram shows a possible minimal interaction vertex for a Higgs boson (always denoted by a dashed line in Feynman diagrams) interacting with a top quark (an example of a massive particle) and its antimatter equivalent:

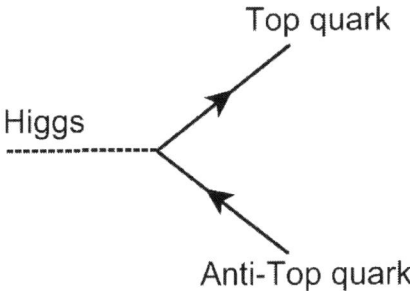

At this point, we have to remember that we can glue together these minimal interaction vertices and the resultant Feynman diagram will still be valid. So we can glue together two of these vertices back-to-back to give the following diagram:

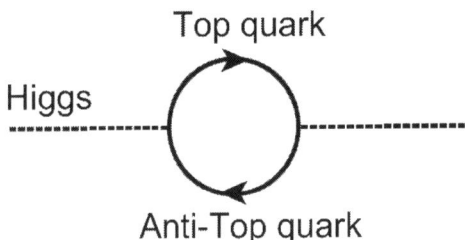

The previous diagram shows a Higgs boson which momentarily converts into a top quark/anti-top quark pair, before converting back to a Higgs boson. A top quark has been deliberately chosen for this example as it is the heaviest elementary particle (i.e., the one with the most mass). And, as this Feynman diagram represents a valid interaction, we have to include it in our calculation of the Higgs' path (remember: Feynman diagrams have to include absolutely anything that can possibly happen). The result is that the weight of the top quark adds to the Higgs mass. Plus, we have to include the effects of all massive particles in our calculations (because the Higgs interacts with all massive particles). The end result is that all these quantum contributions once again push the expected weight of the Higgs up towards the vastly larger Planck mass.

So here we have an apparently huge gap between the expected value of the Higgs mass and the measured value: the Higgs is ten million billion times lighter than we would expect, or sixteen orders of magnitude. This huge gap in the mass values represents a *hierarchy*, and the problem of why the Higgs is so light is called the *hierarchy problem*. The hierarchy problem is considered to be one of the greatest unsolved problems in physics.

The hierarchy problem introduces another fine-tuning problem. If the Higgs mass was lighter, then atoms could not form and life could not exist. But if the Higgs was heavier – and particles were therefore as heavy as the Planck mass – then all particles would turn into black holes (too much mass squeezed into a microscopic volume). Hence, the Higgs mass value appears fine-tuned for the existence of life.

There is a popular hypothesis in physics called *supersymmetry* which has been around for forty years which has been touted as a potential solution to the apparent fine-tuning of the Higgs mass. Supersymmetry requires the existence of a whole suite of additional undetected particles called *superpartners*, each superpartner particle being matched to an existing known particle. According to supersymmetry, the superpartner of a boson is a fermion, and vice versa. The precise symmetry means that the huge quantum contributions from the known virtual particles would be cancelled by the contributions from the superpartner particles. With the huge contributions eliminated, a light Higgs is left. However, no superpartner particles have been discovered at the LHC, and supersymmetry appears to be in trouble. This is a big deal. Many theorists have spent decades of their lives working on supersymmetry. Also, if supersymmetry crashes and burns, then so does string theory which depends on supersymmetry. This has clearly become a high-stakes game. The failure of the LHC to discover new particles leading to the potential demise of supersymmetry – and the dearth of any replacement explanations – has been variously called the "nightmare scenario", a "crisis in physics", or even "the end of science".

Perhaps a new approach is needed …

7

THE WEAKNESS OF GRAVITY

This is the final chapter of this book.

As regular readers of my books will know, I like to include some of my own original ideas at the end of my books – and this final chapter is no exception. So please bear in mind that parts of this chapter should be considered as being highly-speculative.

Although I have tried to present the concepts as clearly and simply as possible, this is still quite a technical chapter.

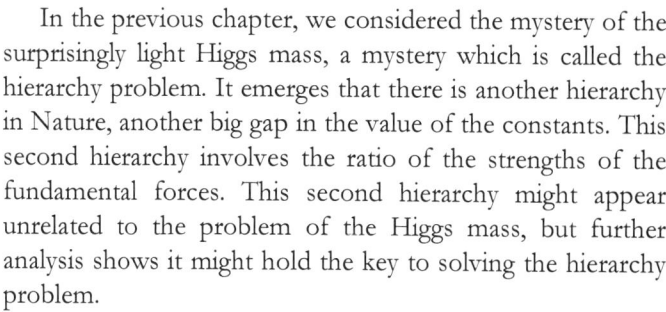

In the previous chapter, we considered the mystery of the surprisingly light Higgs mass, a mystery which is called the hierarchy problem. It emerges that there is another hierarchy in Nature, another big gap in the value of the constants. This second hierarchy involves the ratio of the strengths of the fundamental forces. This second hierarchy might appear unrelated to the problem of the Higgs mass, but further analysis shows it might hold the key to solving the hierarchy problem.

Let us start by reminding ourselves that there are four fundamental forces:

- **Gravity**. Keeps your feet on the ground.

- The **electromagnetic force**, a combination of both the electric force and the magnetic force. In the atom, the electric force attracts negatively-charged electrons to positively-charged protons.

- The **strong force**. In the atomic nucleus, it is the strong force which holds quarks together to form protons and neutrons. It has to be strong to overcome the electrical repulsion between positively-charged protons.

- The **weak force**. In the atomic nucleus, this force can convert a neutron into a proton during radioactive decay.

As explained in my fifth book, in all of these four cases, the actual force is transmitted via gauge bosons, which can be considered as being the "force-carrying" particles. For the electromagnetic force, the gauge boson is the photon. However, as we have seen in the discussion of Feynman diagrams in the previous chapter, the passage of the photon between electrons is constantly disrupted by a swarm of virtual particles. The effect of these virtual particles is to hide the true strength of the electric charge, and to make the force seem weaker over distance than it would otherwise be. This effect is called *screening*. However, if we consider particles interacting at higher energy, these particles will have shorter associated wavelengths and so will be able to "probe" closer to each other. Hence, this will reduce the screening effect and give a truer picture of the strength of the force. And this is what is found through experiment: at higher energies, the electromagnetic force gets stronger.

The opposite is true for the strong force which holds quarks together in the atomic nucleus to form protons and neutrons. Because of the behaviour of the gauge bosons of the strong force (*gluons*), the strong force becomes stronger at large distances. In fact, it is due to the strength of the strong force that no isolated quarks are ever found on their own (it would require so much energy to separate the quarks that new quarks would be produced from that energy).

On that basis, the following graph shows the predicted variation of the strengths of the three forces as the energy of the interactions increases:

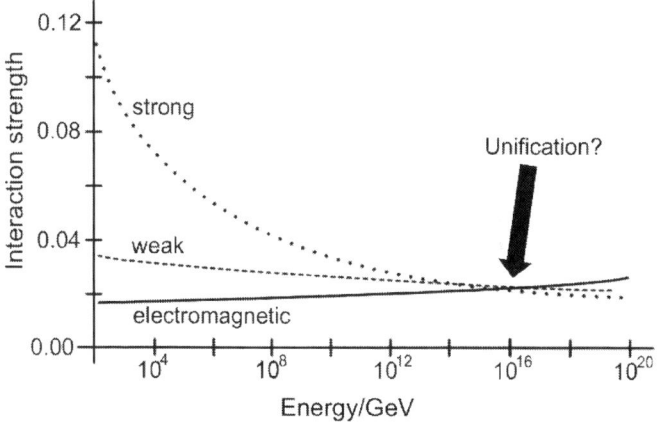

As just described, you can see in the diagram that the electromagnetic force becomes stronger at higher energies, whereas the strong force becomes weaker.

It can be seen that the lines describing the strength of the forces are predicted to approximately intersect at a single point at high energy (marked on the diagram by the "Unification" arrow). You will see from the horizontal scale that this represents an energy of approximately 10^{16} GeV, and this is called the *grand unification energy*. This represents a much higher energy than can be achieved using current particle accelerators. This seems to indicate that the fundamental strengths of all three forces are actually the same, and this would become clear at high energies, but that fact is being hidden by the behaviours of the gauge bosons at low energies.

We now know that the electromagnetic and weak forces have the same origin, and only appear to be of different strengths because of the differences in the associated gauge bosons. The gauge bosons of the weak force have mass (due to the Higgs mechanism) which results in the bosons being short-range and they are therefore confined to the atomic nucleus, whereas the gauge boson of the electromagnetic force (the photon) remains massless and is therefore capable of traversing the universe (bringing light from the stars). The true strength of the weak force is not actually weak at all.

It therefore appears probable that all the forces have fundamentally the same strength. So this makes it all the more mysterious to discover that the strength of the force of gravity is many orders of magnitude weaker than the strengths of the other three forces.

At first glance, it might not appear that gravity is a particularly weak force. After all, it is the force which keeps the planets in orbit around the Sun. It is also the force behind terrifyingly-attractive black holes. However, the planets, and the stars, and black holes need to possess immense mass in order for gravity to become dominant.

To understand the weakness of gravity, consider the act of lifting a metal paperclip by using a magnet. The paperclip is being attracted to the centre of the Earth by the force of gravity, and the entire mass of the Earth is pulling on it. However, the electromagnetic force from the small magnet can easily overcome that pull of the entire mass of the Earth.

So just how weak is gravity?

In the previous chapter we considered the electrical repulsion between two electrons in order to derive the fine structure constant which is the coupling constant which

THE WEAKNESS OF GRAVITY

describes the strength of the electric force. Now let us derive a coupling constant for the gravitational force which describes the strength of the force of gravity.

Let us calculate the coupling constant for gravity in a similar manner to the way we calculated the coupling constant for the electric force. So, instead of starting with Coulomb's law (for the electric force) we will start with Newton's law of gravity. As we considered electrons in the previous example, let us be consistent and consider electrons again (comparing like-with-like). In which case the gravitational force, F, between the two electrons is given by:

$$F = \frac{Gm_e m_e}{r^2}$$

where G is Newton's gravitational constant, and m_e is the mass of an electron (we can obviously replace the two instances of m_e with m_e^2).

Next, as with our earlier derivation of the fine structure constant, we remove the r^2 term which leaves us – once again – with an expression which has dimensions of energy times distance. So we once again have to divide by $\hbar c$. This gives us our final result, a value for the *gravitational coupling constant* which is a dimensionless number expressed purely in terms of fundamental constants:

$$\alpha_G = \frac{Gm_e^2}{\hbar c}$$

So what is the value of this gravitational coupling constant (which describes the strength of the force of gravity)? Well, if you substitute the correct values into the previous formula you will find that the coupling constant for gravity is approximately 10^{-45}, which is obviously an extremely small number. In her book *Warped Passages*, Lisa

Randall refers to this extraordinarily small number when comparing the forces between a pair of electrons: "The gravitational attraction is about a hundred million trillion trillion trillion times weaker than the electrical repulsion between the electron pair."

So here we have another hierarchy, another immensely large number: 10^{45}. Why is there such a huge difference between the strength of the electric force and the strength of gravity? This forms another puzzling hierarchy problem, but it is believed that this particular hierarchy problem is related to the earlier hierarchy problem of the surprisingly-light Higgs mass.

In her book, Lisa Randall considers the hierarchy problem of the Higgs mass and states: "It ultimately boils down to the weakness of gravity compared with all the other known forces." Why is there a connection between the lightness of the Higgs and the weakness of gravity? Well, if you remember back to the discussion of the Higgs mass in the previous chapter, you will remember that the natural value of the Higgs would be expected to be the Planck mass, the formula for which was given earlier as:

$$m_P = \sqrt{\frac{\hbar c}{G}}$$

The hierarchy problem is based on the fact that the Higgs mass is immensely smaller than the Planck mass. Or, to put it another way, the hierarchy problem is therefore based on the fact that the Planck mass is so large. And, for reasons we shall consider shortly, a large Planck mass represents weak gravity. Lisa Randall agrees: "A huge Planck scale mass is equivalent to extremely feeble gravity." So that presents us with a possible opportunity: if we want to solve the hierarchy problem of the Higgs mass, we might only need to solve the problem of why gravity is so weak.

The Randall-Sundrum model

One proposal as to why gravity is weak comes from string theory. String theory requires the existence of many hidden spatial dimensions, and it also predicts the existence of *branes*. Branes are like membranes or sheets in multi-dimensional space, containing fewer dimensions than the space around them. So a brane is a slice of a higher-dimensional world. The full, multi-dimensional world which surrounds the brane is called the *bulk*.

According to string theory, the particle which carries the force of gravity – the *graviton* – is a closed string loop, whereas the particles which carry the other forces are open strings which have to have their ends terminated on a brane. This restricts the motion of those particles to the brane, whereas the graviton is free to travel through the bulk:

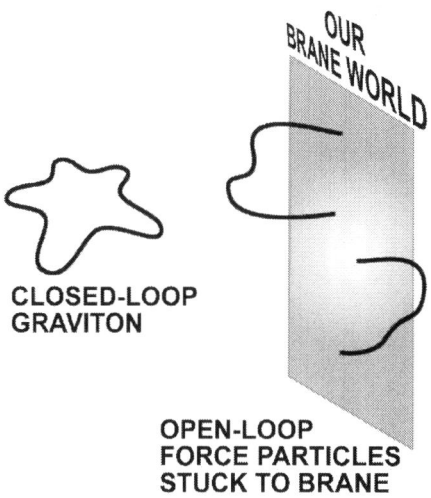

The *Randall-Sundrum model*, proposed by Lisa Randall and Raman Sundrum, is based on the idea that our universe with its three spatial dimensions might be a brane in a higher-dimensional bulk which actually has four spatial dimensions. Unlike the particles associated with the other forces, the graviton is free to travel through the bulk. This, then, presents a proposal for why gravity appears so weak in our universe: we only see a "shadow" of the force of gravity in our three-dimensional universe. The suggestion is that gravity weakens along that fourth spatial dimension as it passes through the bulk to our three-dimensional world:

In her book, which was published in 2005, Lisa Randall listed some of the predictions of the Randall-Sundrum model:

> *Should this warped geometry scenario prove to be a true description of our world, the experimental consequences at the Large Hadron Collider (LHC) at CERN in Switzerland will be tremendous. Signatures of the warped five-dimensional spacetime could include Kaluza-Klein particles, five-dimensional black holes of anti de Sitter space, and TeV-mass strings.*

Any one of these discoveries would have been incredibly exciting. Unfortunately absolutely none of these things were discovered when the LHC started operation in 2010 – although they should have been discovered if the model was correct. Lisa Randall goes on to say: "Results from the LHC are likely to change the way we view the world." Considering the stream of negative results which has emerged from the LHC, I think I can agree with Lisa Randall on that statement – though perhaps not in the way she intended.

Considering the Randall-Sundrum model, imagining an additional invisible dimension of space might represent one route to solving the problem of why the strength of gravity is so weak. But in this series of books I am attempting to build a model of how the universe works in which no unseen extra dimensions are allowed, there are no invisible parallel universes, and you are not allowed to rely on huge suites of undetected supersymmetric particles. In my "reality-based" model, solutions to problems must be found on the basis of what we already know: you are not allowed to invent some parallel universe in which you get to make your own rules.

However, I find those self-imposed constraints concentrate the mind wonderfully. One of the themes of my series of books is that we need to try harder to make sense of what we already know. Sometimes the simplest solution may have been overlooked.

Maybe it is hidden in plain sight?

The true strength of gravity

The weakness of gravity is perceived as being a great mystery, and possibly holds the key to the solution of the hierarchy problem. Let us examine the problem more closely.

Firstly, let us remember that in the previous chapter it was stated that we would naturally expect the mass of the Higgs boson (and therefore the mass of the other particles) to be close to the value of the Planck mass. And the fact that the Higgs is orders of magnitude lighter than this expected value forms the hierarchy problem. But why should we expect the masses of the particles to be naturally close to the Planck mass?

Well, to answer that question, let us remind ourselves of the formula for the Planck mass:

$$m_P = \sqrt{\frac{\hbar c}{G}}$$

and also let us remind ourselves of the formula for the gravitational coupling constant which we derived earlier (remember, the small value of this expression is the reason for weak gravity):

$$\alpha_G = \frac{Gm_e^2}{\hbar c}$$

You will see that the mass of the electron features in this formula for the gravitational coupling constant. But what happens if the electron mass was instead equal to the Planck mass? What would be the strength of gravity in that case?

THE WEAKNESS OF GRAVITY

To answer that question, let us substitute the value of the Planck mass into the previous formula for the gravitational coupling constant. That gives us:

$$\alpha_G = \frac{G \times \left(\dfrac{\hbar c}{G}\right)}{\hbar c} = \frac{\hbar c}{\hbar c} = 1$$

So, you can see from this formula that the value for the gravitational coupling constant (and, therefore, the strength of gravity) would be equal to 1 for a particle whose mass is equal to the Planck mass.

(Conversely, if the mass of a particle is far less than the Planck mass then it will experience weak gravity – which is the reason why the large Planck mass represents a weak force of gravity in the universe).

At what energy will the strength of gravity become equal to the strengths of the other forces? Well, if you substitute the value of the Planck mass into $E=mc^2$ you will discover it represents an energy of approximately 10^{19} GeV. This is called the *Planck energy*. This is a huge amount of energy for a single particle, approximately equal to the chemical energy stored in a car's petrol tank. You will see that this value is only moderately greater than the value of the grand unification energy (10^{16} GeV) which we considered earlier. This seems to suggest that for particles with extremely high energies – just beyond the grand unification energy – the strengths of all four forces would be approximately equal.

It therefore seems likely that the true strength of gravity is the same as the other three forces, and this would become clear at high energies. I believe this is fairly generally accepted. This is a valuable insight we can use in our quest to discover why gravity now appears so weak.

Now let us continue by considering a vital fact about the laws of physics.

The laws of physics do not change over time

Emmy Noether was a mathematician who has been described as "the most important woman in the history of mathematics". Emmy Noether worked in academia in the early decades of the 20th century, which was not an easy time to be a female scientist. However, she showed great determination and resilience to make several important contributions to mathematics, and also make an extremely important and influential contribution to theoretical physics.

Noether's great contribution to physics is now called *Noether's theorem*. Put simply, Noether's theorem states that if there is any symmetry in Nature then there will also be a corresponding conserved quantity, i.e., a quantity which does not change over time. Noether's theorem was considered in detail in my fifth book, which explained the tremendous value of the theorem to particle physics.

In that book, an example was presented of the application of Noether's theorem to a skateboarder riding in a halfpipe. It was explained that if there was a symmetry in space, i.e., if the skateboarder was riding down the length of the halfpipe, then the quantity that would be conserved would be the momentum of the skateboarder. In other words, the skateboarder would continue to travel at a constant speed.

So when there is a symmetry in space, momentum is conserved.

Noether's theorem indicates that there is another similar relation between time and energy: if there is a symmetry in time, energy is conserved. But what does symmetry in time actually represent? Well, it means that if we move some physical process – maybe an experiment – along the time dimension then we would find that the underlying situation would be unchanged (that is the definition of a symmetry). As an example, if I perform an experiment today and get a certain result, then if I perform precisely the same experiment again tomorrow then I will get exactly the same result. So what symmetry in time really means is that **the laws of physics do not change over time.**

Noether's theorem tells us that symmetry in time implies conservation of energy. Is there any easy way of understanding why this should be the case? Well, imagine if the strength of gravity (one of the laws of physics) varied with time. In that case, when gravity was weak I could lift a heavy object up some pulley system, connected to an electrical generator. Then when gravity became strong again I could release the weight and get back more energy (in the form of electrical energy) than I put into the system. I could get free energy! This obviously cannot be the case, so the laws of physics not changing with time implies conservation of energy.

I believe that the fact that the laws of physics do not change with time plays a key role to understanding why the strength of gravity is weak. To see why that is the case, let us again consider Newton's law of gravity, which is obviously one of the laws of physics and which should therefore not alter with time:

$$F = \frac{Gm_1m_2}{r^2}$$

The r^2 term represents an *inverse square law* which means the force reduces according to the square of the distance. This is an inevitable consequence of the three dimensions of space (as described in my previous book). It is clear that this aspect of the equation would not vary with time.

However, you will notice that the equation also includes Newton's gravitational constant, G. If the laws of physics do not change with time then that implies that **the value of G**

does not change with time. I believe this is an important result in explaining why the strength of gravity now appears so weak, as we shall soon see.

You might not think it is such a big deal to state that the value of G does not change with time. After all, it is supposed to be a fundamental constant! However, that has not stopped the emergence of several theories which suggest a varying value of G. Perhaps the most famous example is that of Paul Dirac, one of the greatest physicists of the 20th century. In one of his wilder flights of fancy, Dirac proposed that the huge ratio of the strength of the electromagnetic force to the strength of gravity was equal to the ratio of the radius of the universe to the radius of a proton. Unfortunately, in an expanding universe, this would mean that both ratios would have to be smaller in the past than they are is now. How could that possibly be the case if the strengths of the forces does not change? The only way Dirac could think of avoiding this problem was to suggest that the value of G was actually larger in the past (thus reducing the ratio in the past). Could it really be the case that G has varied with time? Dirac stuck doggedly with his theory, but it was shown that if it was the case that G was larger in the past then this would have resulted in increased energy output from the Sun: the oceans would have boiled and life could not have survived.

I firmly believe that we should take it as a rock-solid fact that G has not varied with time. I have always taken it as one of my guiding principles. In fact, I would suggest that if you have a theory which predicts a varying G, you would be wise to throw your theory in the bin.

The early universe

For our next step in trying to construct a potential solution to why gravity is so weak, we will need to travel back in time. A long way back in time.

Physicists now have a very accurate model of the early universe right back to a tiny fraction of a second after the Big Bang. This was a time of extremes, of tremendous temperatures and pressures. The temperature of the universe at this point in time is believed to have been the highest possible temperature which is known as the *Planck temperature*. The Planck temperature is equal to 10^{32} K, or 100 million million million million million degrees kelvin. Heat is the result of particles moving randomly at tremendous speed. This means particles had immense kinetic energy.

We can understand this situation by considering a large number of particles (for example, a gas) trapped in a box. The box is then heated to a great temperature. Heat is the result of particles moving randomly at tremendous speed, so the kinetic energy of the particles increases. As a result, the total energy of the box increases, and so does its mass. As the Wikipedia page on "Mass in Special Relativity" explains:[14]

---

[14] http://tinyurl.com/massinrelativity

If a stationary box contains many particles, it weighs more in its rest frame, the faster the particles are moving. Any energy in the box (including the kinetic energy of the particles) adds to the mass, so that the relative motion of the particles contributes to the mass of the box.

And what is the universe if not a big box full of particles?

As you will see in the following diagram, the Planck temperature of the early universe corresponds to the average energy of particles being 10^{19} GeV, a value which we calculated earlier as being the Planck energy. And, as we calculated earlier, this is the immensely high energy at which the strength of gravity becomes equal to the strengths of the other three forces.

As a result, it is suspected that all the forces were unified into a single force at the time of the Big Bang, and separated into individual distinct forces as the universe cooled.

The following diagram shows this process. Time is on the horizontal scale, with the Big Bang on the left of the diagram:

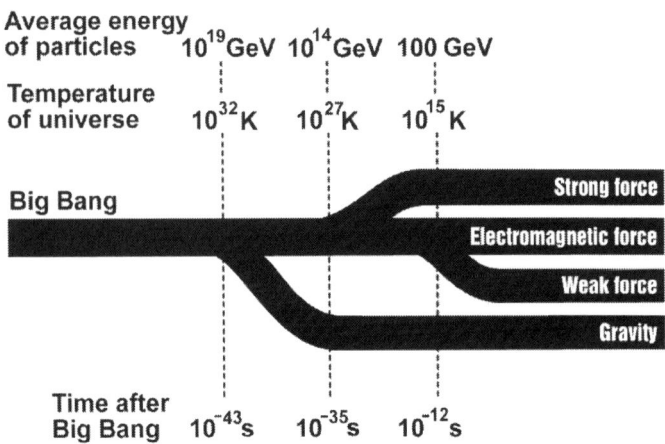

THE WEAKNESS OF GRAVITY

You will see that the first force to split into a separate distinct force was the force of gravity – which is what interests us. You will see from the diagram that this happened just a fraction of a second after the Big Bang: 10^{-43} seconds, to be precise, an incredibly short period of time which is known as the *Planck time* (the period before this time is known as the *Planck era* or *Planck epoch*). Due to quantum uncertainty, we seem to be fundamentally limited about what we can know about conditions during the Planck era.

As discussed earlier, it is believed that 10^{16} GeV represents the grand unification energy at which the three remaining forces (strong, weak, and electromagnetic) are still unified, but as the universe cooled beyond that point, those three forces separated into distinct forces.

As the universe continued to cool, you can see that at the much lower energy of 100 GeV, the electromagnetic and weak forces split from the previous unified *electroweak* force. This was due to *spontaneous symmetry breaking* by the Higgs field, giving mass to the bosons of the weak force while leaving the photon massless. This energy of 100 GeV is low enough to come within the energy range of our current particle accelerators. And, indeed, the LHC recently detected the Higgs boson (the particle associated with the Higgs field) with a mass of 125 GeV.

Which places us in a position to attempt to answer the main question ...

Why is gravity so weak?

The extreme weakness of gravity is presented as one of the great mysteries of physics. In fact, it has even been called "the greatest unsolved problem in theoretical physics".[15] Many articles have been written on the subject, and several models have been produced such as the Randall-Sundrum model in order to explain the phenomenon. These models have even incorporated highly-speculative extra dimensions – all with the aim of explaining the weakness of gravity.

However, there does appear to be a very simple solution – an almost trivially simple solution – to this question which only involves conventional orthodox physics. In order to explain the solution, let us return to consider the earliest moments in the life of the universe.

We know that the force of gravity split to become a distinct force in just a fraction of a second after the Big Bang. If there was a force of gravity at that point in time, then there must also have been a law of gravity. And, as we know the laws of physics do not change with time, that law of gravity must have been the same as it is now: Newton's law of gravity must have come into existence 10^{-43} seconds after the Big Bang. So the value of Newton's gravitational constant, G, must have been fixed just 10^{-43} seconds after the Big Bang. It is as if a snapshot was taken of G at that time, and we have been using that value ever since:

---

[15] Ethan Siegel, *The Greatest Unsolved Problem in Theoretical Physics*, http://tinyurl.com/greatestunsolved

THE WEAKNESS OF GRAVITY

For the next step, we need to realise that this means that the value of *G* would have been fixed **based on the extreme conditions prevailing at that time.** At that early moment in time, the energy of particles was equal to the immense Planck energy (you can see this on the earlier diagram showing the progressive separation of the four forces after the Big Bang: "Average energy of particles: 10^{19} GeV"). The value of *G* would therefore have been set on the basis that the average energy of particles was far greater than today. And, because of the increased energy of those particles, this would indeed have resulted in gravity having the same strength as the other three forces at that early moment in time (because, as we calculated earlier in this chapter, even today gravity would have the same strength as the other forces if the energy of particles was equal to the Planck energy).

This makes sense in another way as well because it means the natural, expected value for mass at that time could be calculated in the usual manner from the three most basic fundamental constants, without including any arbitrary constants, which would be equal to the Planck mass:

$$m_P = \sqrt{\frac{\hbar c}{G}}$$

And the Planck mass was, indeed, the average, expected value of the mass of particles at that early time in those extreme conditions, when the laws of physics were set in stone.[16] And it has remained the "natural" value for mass ever since – even though much has changed in the universe since then …

Over the period of time since the Big Bang, the universe has expanded and cooled to a remarkable degree. From that initial temperature of 100 million million million million million degrees, within three minutes it had rapidly cooled to a positively chilly billion degrees. And with that rapid fall in temperature came a corresponding reduction in the kinetic energy of particles. Remember, the kinetic energy of particles dominated in the early universe, but with the average temperature of the universe now at merely a couple of degrees kelvin, the kinetic energy of particles currently forms a negligible part of their total energy.

Which brings us to the next crucial point. Mass and energy can be considered to be the "charge" of gravity: the more mass and energy a particle has, the more it feels the force of gravity. So the cooling of the universe – and the corresponding reduction in energy – has implications for gravitational charge. **But this is not the case for the other forces.** Electric charge, for example, does not vary with temperature. As described at the start of this chapter, there is

---

[16] With the Planck mass being calculated from the Planck energy. This refers to the *relativistic mass* of particles, as opposed to their *rest mass*. If the particle was placed on weighing scales, it would weigh more due to the increase in its relativistic mass. Relativistic mass includes the contributions from both mass **and** energy, which makes it more important as far as gravity – and weighing scales – is concerned (mass and energy are the "charge" of gravity).

THE WEAKNESS OF GRAVITY

a minor variation in the strength of the electromagnetic force with energy due to the screening effect, and a graph was presented showing how the strengths of the three forces varies with energy. But this variation would be negligible compared to the variation in the strength of gravity. The cooling of the universe would therefore have left the strengths of the other three forces unchanged – it is the strength of gravity alone which would be affected by many orders of magnitude.

To sum up, the law of gravity uses a gravitational constant, G, the value of which was set when the average gravitational charge was huge. After the cooling of the universe, the average gravitational charge was vastly reduced. With the strengths of the other forces unaltered, gravity now appears relatively weak.

In the discussion on naturalness in Chapter Four, it was explained how theories have a major challenge in explaining the origin of any constant which takes the form of a large dimensionless number. What is the origin of the large number? What mathematics could possibly generate it? Any theory which attempts to explain the weakness of gravity must explain the origin of the large dimensionless number which describes the ratio of the strength of the other forces to the strength of gravity. As I say, it is a major challenge for any theory to identify the origin of this unnaturally-large dimensionless constant.[17] However, in this discussion <u>considering the relative weak</u>ness of gravity, a simple

---

[17] The Randall-Sundrum model for weak gravity gets its immense number from the *exponential* weakening of gravity as it moves from the gravity brane to our three-dimensional world. As Lisa Randall says: "An exponential automatically turns a modest number into an extremely huge number."

explanation of the origin of that particular immense number has emerged: the origin is the ratio between the immense temperature of the universe at the time of the Big Bang, and the cool average temperature of the universe now. In other words, one immense number has been converted into another immense number.

On that basis, a simple calculation can be performed to quantify the effect of the cooling universe on the relative weakness of gravity. Energy is proportional to temperature.[18] So to understand the drop in energy, we only need to consider the drop in temperature. As described earlier, the temperature just after the Big Bang was 10^{32} K (the Planck temperature). With the current temperature of the universe being a mere 2.73 kelvin, that represents a dimensionless ratio of temperatures of approximately 10^{32}. This value of 10^{32} representing the relative weakness of gravity is very much the kind of value we are looking for, coming right in the middle of the following range:

- 10^{45} – the strength of the electromagnetic force relative to the strength of gravity.

- 10^{24} – the strength of the weak force relative to the strength of gravity.

To put it simply, it appears that the reason why gravity is now so weak is that gravity was designed for a much heavier world.

---

[18] Energy is equal to kT, where k is *Boltzmann's constant*, and T is the temperature. You can also see that energy is proportional to temperature from the energy and temperature values on the earlier diagram showing the progressive separation of the four forces after the Big Bang.

The fine-tuned universe

I hope you have enjoyed this book.

The intention has been to show that grounded, conventional physics provides the best route to solving the apparent fine-tuning problems. Other techniques – such as the strong anthropic principle – might appear to provide shortcuts to quick solutions, but the temptation to go down those routes should be resisted.

In trying to discover solutions to the puzzling fine-tuning problems, it seems as though whenever progress has stalled, the common approach has been to introduce some additional undetected structure as a way of circumventing the problem. Structures which have been invented to solve the fine-tuning problems include:

- **Supersymmetry**. The proposal for a huge suite of undetected supersymmetric particles as a way of explaining the hierarchy problem. None of these particles were detected by the LHC, which is currently our highest-energy particle accelerator. Unfortunately, as these particles have gone undetected as accelerator energies have progressively increased, some physicists have chosen to conveniently increase the predicted masses of these particles so that they lie just beyond current accelerator energies. In other words, the predictions have been modified in order to preserve the theory. That is surely not a good way to do science.

- **Extra dimensions**. The proposal for unseen and undetectable extra spatial dimensions as a means for explaining the weakness of gravity.

- **The inflaton field**. An additional unexplained and undetected field required by the inflation hypothesis, as a means for explaining the fine-tuned flat universe.

- **The multiverse**. An undetected and undetectable group of universes beyond our own universe, used by the strong anthropic principle to potentially "solve" (in the loosest sense of the word) **all** of the puzzling fine-tuning problems.

Can we really continue proposing new structures whenever we appear to hit a dead-end? Is this really the best way to solve the fine-tuning problems? The danger is, of course, that we are trying to solve the fine-tuning problems by inventing structures that simply do not exist.

Surely it would be better if the process of inventing and proposing new undetected structures was treated as a last resort, when all else has failed. Instead, it seems as if this approach is being treated as the preferred first option.

The message of this book has been that we should try harder to make sense of current data using our existing tools and knowledge – rather than inventing additional unseen structures.

The solutions we seek might well be hidden in plain sight.

The principle of naturalness appears to be a useful guide in our efforts to describe fundamental mechanisms. In an entirely natural solution we would only be left with dimensionless constants of the order of magnitude of one. We would then be entitled to consider the value of that constant as being "natural", and not fine-tuned. We may not know the precise origin of the value of the constant, but we can be satisfied we are not far away from knowing the truth. Examples of "natural" constants might include the fine-structure constant, and the number of dimensions of space. I suspect it is likely that simple solutions to these problems will arise at some point. Maybe we don't have all the answers yet, but we can afford to be hopeful in these cases.

In a completely "natural" universe there would be no arbitrariness whatsoever: the state of that universe would be uniquely fixed as a logical necessity. However, there seems to be a fairly widespread belief that the state of our universe could not entirely arise as a logical necessity, in which case – despite our best efforts – we might be forever left with the question of why the universe takes the form it does. In that case, the only answer to why the constants take the values they have would be "that's just the way it is". I certainly do not believe we have reached that point yet, but if we do reach that point then we would be at the limit of possible human knowledge (for example, quantum uncertainty limits our knowledge of conditions during the Planck era).

The discussion of the Fermi paradox in this book seemed to suggest that life is a rare occurrence in the universe, requiring an almost impossibly special combination of conditions – such as those found on Earth. It has been said that the Earth appears to be a "gem". So maybe we should view this as an opportunity. If we eventually come to the limit of possible human knowledge, we should not despair about the "end of science". Instead, we should congratulate ourselves on a job well done, and spend our time enjoying our lives in this amazing universe on this very special planet.

EPILOGUE

Regarding the hypothesis of modified gravity I presented in my earlier books, there has been a recent development which is potentially very interesting. The modified gravity hypothesis did not only predict a naturally-flat universe, but it also predicted a modification to the structure of black holes. The hypothesis eliminated the troublesome singularities which are supposed to lie at the heart of black holes. Instead, the hypothesis predicted that the mass of a black hole would be concentrated at its event horizon.

In August 2016, a team led by Vitor Cardoso of the Superior Technical Institute in Lisbon suggested that the signature of a thickened event horizon could be detected in the recently-discovered gravitational wave signals. The gravitational waves were produced by the merging of two black holes, producing a staggering amount of energy. This prediction was investigated by a team led by Niyayesh Afshordi of the Perimeter Institute and the University of Waterloo.

In December 2016, Afshordi's team released a paper with the evocative title of "Echoes from the Abyss", in which they announced that they had detected echoes of energy released between the inner layer of the event horizon and the outer layer of the event horizon. This would appear to indicate some structure at the event horizon, structure which was not predicted by conventional general relativity. The announcement was also featured in an article in *Nature* magazine which was aimed at the general readership.[19]

Being realistic, the chances that this result will prove to be conclusive evidence of a modification to general relativity is extremely remote. However, what is exciting is that new data is starting to pour in, and will continue to pour in for the next few decades as a series of new ambitious experiments go online. The next few years might prove to be very exciting indeed.

Watch this space!

[19] *LIGO black hole echoes hint at general relativity breakdown*, http://tinyurl.com/blackholeechoes

FURTHER READING

Just Six Numbers by Martin Rees
An analysis of six dimensionless constants which appear to be set to life-friendly values.

The Theory That Would Not Die by Sharon Bertsch McGrayne
The fascinating story of Bayes' theorem. The paperback edition includes technical details and worked examples.

The Copernicus Complex by Caleb Scharf
Are we alone in the universe? This very readable book comes to a stunning conclusion.

The Eerie Silence by Paul Davies
The search for extraterrestrial intelligence.

QED: The Strange Theory of Light and Matter by Richard Feynman
An accessible introduction to quantum electrodynamics.

Feynman by Jim Ottaviani and Leland Myrick
A delightful comic-strip account of the life of Richard Feynman, including good science.

Let's Draw Feynman Diagrams! by Flip Taneda
http://tinyurl.com/fliptaneda
An excellent resource for the details of Feynman diagrams and the Higgs boson.

PICTURE CREDITS

All photographs are public domain unless otherwise stated.

Enigma machine photograph is courtesy of the Bundesarchiv. Enigma wheels diagram is based on a diagram by Messer Worland and is provided by Wikimedia Commons.

Photograph of Frank Drake is courtesy of the SETI Institute.

Photograph of Curiosity rover and Exoplanet Travel Bureau poster is courtesy of NASA.

Proxima b artist impression is courtesy of the European Southern Observatory ESO/M. Kornmesser.

Solar sail artist's impression is by Kevin Gill and is provided by Wikimedia Commons.

Edwin Hubble's graph is courtesy of NASA.

Drawing of Feynman path integral is by Matt McIrvin and is provided by Wikimedia Commons.

HIDDEN IN PLAIN SIGHT 8

How to make an atomic bomb

COMING SOON

Made in the USA
Lexington, KY
07 July 2017